ALIEN ENCOUNTERS

ALIEN ENCOUNTERS

The Deception Menace

JAMES L. THOMPSON

First Printing: August 1995

International Standard Book Number:
0-88290-536-8

Horizon Publishers' Catalog and Order Number
1058

Printed and distributed
in the United States of America by

Horizon
Publishers
& Distributors, Incorporated
P.O. Box 490 Bountiful, Utah 84011-0490

Contents

1 An "Epidemic" of Alien Encounters? 11

A Common Introduction 11
The 1991 Roper Organization Survey 12
A New Cultural Marvel 18
Uncovering The Deception 20
The Approach Taken In This Book 23

2 UFOs In Antiquity 25

UFO Reports Throughout The Ancient World 25
Scriptural Accounts? 25
Chariots Of Fire 26
The Camp Of Israel Led By A UFO 26
The Wheels Of Ezekiel 28
Von Daniken's "God And Angels Were Aliens" Theory 31
Enoch's Escort 33
UFOs In Egyptian Religious Texts 34
"Alien" Encounters Throughout History? 35

3 Modern UFO Sightings 37

UFO Reports In Recent History 37
UFO Expectations 37
UFOs In The Twentieth Century 39
Public UFO Reports By Credible Witnesses 39
The Modern Era Of UFOs 40
The Controversy Over UFO Evidence 42
UFO Photographs 42
The Gulf Breeze UFOs 44
UFO Design 45
The Blue Beam 47
Telepathic Communication 49
Government Coverup—Men In Black 52
UFOs Around Energy Installations 53
Gentle Invaders 54
From Hardware To Soft Sightings 54
Spherical Probes 55

Modern Airships 56
Mother Ships 58
4 Saucers On The Ground 62
Landing Sites 62
Crash An Retrieval Coverups 63
The Rosewell Crash 65
Area 51 66
More Government Coverups 69
A Documentary Smoking Gun 69
Government Disinformation? 70
If There Is A Coverup—Why SETI? 72
Invisible UFOs 73
The Tunguska Event 73
5 Extraterrestrial Visitors 75
The Nature Of UFO Occupants 75
The Subterranean Civilizations Theory 75
The Returning Earthling Theory 76
The Extraterrestrial Human Theory 77
Small Greys 77
The Varied And Ephemeral Nature Of UFO Occupants 79
Shootout At Kelley 80
Udo Wartena 81
6 Extraterrestrial Abductors 84
Missing Time 84
Judi Moudy 84
The Betty And Barney Hill Abduction 87
Onboard Medical Examinations 89
Antonio Villas-Boas 90
Human Accomplices 91
The Nursery 91
The "Aliens" Have Something To Learn From Us 93
Erased Memories 94
Hypnotic Retrieval 94
The Ed Walters Abductions 95
Implanted Monitors 96

Dreams Or Memories? 97
The Bedroom Encounter 98
Underwater Joyrides 99
Travis Walton 100

7 Strange Encounters 104

The Mojave Triangle 104
Eddy Page 116

8 Contact 124

Early Contactees 124
Public Rejection of Early Contactees 127
Contactees—The Next Generation 128
The One 130
Follow The Son 134
Year 2000 137
Salvation 139
The Message 144

9 What Is The New Age? 146

A Brief History Of The New Age 146
Who Makes Up The New Age Movement? 147
What Do New Agers Believe? 147
New Age Gurus And God 148
When And How Will The New Age Arrive? 149
The New Age Movement Is Pervasive 151
Earth Spirit 153
A UFO/New Age Connection 154

10 The Western View Of The New Age 156

God's Plan 156
Life After Life 160
Reincarnation 161
Ministering Spirits 162
The New Age Will Bring Antichrist To Power 165

11 Paranormal Properties Of UFOs And Aliens 167

UFOs Are Not What They Appear To Be 167
The Healing Of Dr. "X" 169
UFO And OBEs 172

UFO And ESP ... 174
The Entity Enigma ... 175
Reality Testing ... 176
Historical Parallels ... 176
The Secret Commonwealth ... 177
Magic Wands ... 179
Crossing The Time Zone ... 180
Elves And Fairies ... 181
Modern Gentry ... 182
Catholic UFOs ... 184
Blessed UFOs ... 185
Ouija Boards And Seance Rappings ... 186
Photographing The Unknown ... 189
Ghostly Apparitions ... 190
Past Life Recall ... 191
Panoply Of The Paranormal ... 196

12 Who Is Sending The Message? ... 198

Isn't It Obvious? ... 198
A Paradigm For A Plethora ... 198
Theory 1: Extraterrestrial Explorers ... 198
Hard Evidence ... 199
Photographs ... 199
Prime Time Videos ... 200
Target Practice ... 202
Personal Testimony ... 202
The Personal Touch ... 202
Anything Is Possible ... 204
Theory 2: Returning Earthlings ... 204
Theory 3: Natural Phenomena ... 205

13 Are UFOs And The Alien Message A Satanic Deception? 207

Counterfeits Are Patterned After Truth ... 207
Power Of The Air ... 208
Deceptive Devices ... 209
Generated Images ... 209
Spiritual Projections ... 210
Physical Spirits ... 213

Solid Objects Without Mass 216
Synchronous Specters 217
I Am Legion 217
Ghostly Abilities 220
Invasion Of The Body Snatchers 220
Theory 1: Genetic Engineering 221
Why Hybrids? 221
Not To Preserve The Human Form 222
The Superhumans 224
Satanic Science 226
Theory 2: Mechanical Invention 227
Similar Encounters Have Historical Precedents 230
Theory 3: Corporeal Possession 232
Mystic Knowledge 235
Whose Will? 238
Many Indicators Of Corporeal Possession 242
Hypnosis—A Spiritual Revolving Door? 243
Spiritual Psychology 244
The Control System 246
New Ideas 248
The Cosmic Gospel 251
Conclusion 255
Index 258

1
An "Epidemic" of Alien Encounters?

A Common Introduction

Stambul, an acquaintance of mine, had invited a group of us over to his home to see some Unidentified Flying Objects (UFOs). He asked us to arrive just before dusk—dusk was the best time to see UFOs from his balcony. Bolzano, where I was living in 1979, is a beautiful Germanic town in Northern Italy, just south of the Austrian Border. It is thought by its citizens to be a UFO center—many of them claim to have seen a UFO or two in their day. In fact, this region of Northern Italy is known to have had a wave of UFO sightings during this period. I briefly discussed Stambul's invitation that day with an American colleague of mine, Stan, and we decided to take Stambul up on his offer, but the topic did not occupy too much of our attention for the remainder of that day.

At the predetermined rendezvous time, Stan and I, along with a couple of others, were punctual, and there was just enough time to exchange a few pleasantries and consume some Italian "biscotti" before Stambul suggested we should go out on the veranda. The small group stepped out into the evening air. The city lay still in the eventide shadows of the eclipsing Dolomite mountains towering around us. We spoke in hushed tones, politely scanning the darkening sky—watching for UFOs.

Two minutes had not passed when Stambul spoke in his casual Italian voice, "Oh look, there's one now," pointing above us. We looked up, and watched as a pinpoint of light shimmered its way slowly through the twilight. I surveyed the object as it glimmered, moving in an almost straight path over us, quivering slightly every five seconds or so as if it had hit some ice and veered from its course momentarily. My quiet observation was disturbed by the commotion of others in our group, pointing and exclaiming. I looked at Stan and exchanged a smile—it looked like a passing satellite to us.

Another of Stambul's "UFOs" had flown over within five minutes, and we thanked him as we departed, suggesting that although the objects may have been satellites, who could tell for sure?

Later that evening, Stan solemnly related to me his personal encounter with a UFO. He was a well-grounded, solid, conservative young man, now a successful business attorney, and I knew that his report was true. A couple of years ago, Stan was kind enough to share the following narration of the encounter.

> When I was still in high school (mid-1970s) I was dog-sitting for a family acquaintance. She was completely blind and the dog was her seeing-eye dog. To keep the dog in peak performance (a seeing-eye dog that is not in peak performance is a sorry sight to behold), it was necessary for me to take the dog on a rather long walk and put it through a series of exercises each day. Demands being what they were in those days (some things never change), the only time I could engage in such activities was at night. It was while I was on one of these walks that I had my UFO experience.
>
> On that particular night we had walked to an area where there was a lot of new construction but very few occupied structures. There were no artificial lights to obstruct the view. The sky was clear and afforded quite a spectacular view. As I recall, my mind was not on the view, however. In fact, I believe I had my eyes closed at the time trying to get the full impact of being guided by a seeing eye dog. The dog came to a sudden stop which caused me to open my eyes. I followed the dog's eyes and saw a saucer-like object hovering in the air. The craft had fixed lights on the top and the bottom with rotating lights around its perimeter. In the desert air at night, sounds carry for a great distance. The craft, however, did not emit any audible sound.
>
> The craft continued to hover in a uniform pattern for a period of about 30 seconds, then it made a very quick movement to one side, returned to what seemed to be the original location and made a very quick vertical move, disappearing from my view entirely. The dog continued to watch the sky for a moment and then continued on as though nothing had happened.

Stan's sighting was the second such encounter that had come to my attention, of someone I knew well and whose word I trusted. It caused me to wonder for the next dozen years whether or not there could be any substance to the UFO-encounter phenomenon.

The 1991 Roper Organization Survey

The Roper Organization was commissioned in 1991 to conduct a survey of American adults to determine the extent of American participation in the

UFO/alien encounter phenomenon.[1] The survey was commissioned by the Bigelow Holding Company, a consortium of UFO investigative groups, scientists, and other interested persons, for the purpose of alerting the mental health profession that patients with claims of close encounters and abduction reports are not necessarily delusional. The results of the survey were published in late 1992, along with a plea to the mental health community by John E. Mack, M.D., Professor of Psychiatry at The Cambridge Hospital, Harvard Medical School. As of this printing, the survey results were still sending shock waves through the mental health profession, as well as the media.

Since the publication of the survey results, Dr. Mack, also the Pulitzer Prize-winning author of *A Prince of Our Disorder* (a biography of T.E. Lawrence), has published his own book on the alien encounter/abduction phenomenon titled *Abduction: Human Encounters with Aliens* (Charles Scribner's Sons, New York: 1994). Since being introduced to the encounter /abduction phenomenon about four years earlier by famed abduction researcher, Budd Hopkins, Dr. Mack has become quite enamored of the topic. He has devoted many of his professional resources to investigating "alien abductions."

Three Roper "Omnibus" polls (July, August, September of 1991) involving a total of 5,947 adults, repeated a battery of eleven questions, five of which were key indicator questions. Some questions were pointed, and some subtle, while one was a mere control question inserted to determine how many "yea-sayers" were participating. Surprisingly, only one percent responded affirmatively to the control question, "Hearing or seeing the word TRONDANT and knowing that it has a secret meaning for you." The pollsters contrived the word Trondant.

The survey was conducted so scientifically that the results are given a margin of error of only plus or minus 1.4 percent. The most pointed question was "Have you seen a UFO?" The response was a surprising 7 percent of all American adults.[2] That is 7 percent of 185 million,[3] or a total of 12.95 million adults.

---

1 *Unusual Personal Experiences: An analysis of the Data from Three National Surveys,* The Bigelow Holding Company, Nevada (1992).

2 *Ibid.*, p. 7.

3 This survey excluded all children, the populations of Hawaii and Alaska and anyone living in shared, institutional quarters. *Ibid.*, p. 48.

In all surveys, there is a segment of the adult population known as the Political Social Actives (PSAs), which is made up of those who actively participate more actively in the community and social affairs. This participation can be with the PTA, political parties, or with other organizational efforts—generally thought of as trend setters rather than trend followers. This PSA segment of society essentially remains a constant 10 percent of the general population. What is interesting is that PSAs responded affirmatively to the survey questions around 40 to 50 percent more often than the average American adult. To the above question, regarding having seen a UFO, the PSA response was 10 percent.[4]

The survey asked if the participant had experienced the following, which are indicators (in the opinion of the commissioners) of a prior UFO abduction. Included are the questions and the responses, both average and PSA. Take the survey yourself before being exposed to the information in this book, and see how your own experiences compare to those of other American adults:

> This card contains a list of things that might have happened to you at some point in your life, either as a child or as an adult, or both. I'd like you to read down the card, and for each item tell me, to the best of your knowledge, if that has happened to you more than twice, once or twice, or never.
>
> a. Seeing a ghost [Av 11% PSA 16%]
>
> b. Feeling as if you left your body [Av 14% PSA 23%]
>
> c. Seeing a UFO [Av 7% PSA 10%]
>
> d. Waking up paralyzed with a sense of a strange person or presence or something else in the room [Av 18% PSA 28%]*
>
> e. Feeling that you were actually flying through the air although you didn't know why or how [Av 10% PSA 18%]*
>
> f. Hearing or seeing the word TRONDANT and knowing that it has a secret meaning for you [Av 1% PSA 2%]
>
> g. Experiencing a period of time of an hour or more, in which you were apparently lost, but you could not remember why, or where you had been [Av 13% PSA 17%]*
>
> h. Seen unusual lights or balls of light in a room without knowing what was causing them, or where they came from [Av 8% PSA 11%]*

---

4 *Ibid.*, p. 31.

i. Finding puzzling scars on your body and neither you nor anyone else remembering how you received them or where you got them [Av 8% PSA 9%]*

j. Having seen, either as a child or an adult, a terrifying figure—which might have been a monster, a witch, a devil, or some other evil figure in your bedroom, closet, or somewhere else [Av 15% PSA 19%]

k. Having vivid dreams about UFOs [Av 7% PSA 10%]

To the uninitiated, many of these questions may appear to be unrelated to the UFO phenomenon. What do bedroom visitations, unnoticed scars, and unusual lights in the house have to do with UFOs? Many would associate a number of these questions with spiritual or psychological manifestations that should be dealt with in those terms. The commissioners of the survey concluded that, based upon their inside knowledge of the UFO phenomenon, anyone who answered affirmatively to four out of the five questions followed by "*" is likely to have been abducted by aliens without his or her knowledge. Is this a credible conclusion for scientists and doctors to reach? You be the judge as you examine the materials included herein.

The Roper Organization's survey produced results that were unexpected by the survey's commissioners.[5] They had anticipated much lower levels of persons reporting that they had actually encountered the phenomena enumerated in the survey. Moreover, approximately 30 to 40 percent of those responding in the affirmative indicated that they had experienced the various phenomena on more than one occasion.

In converting the responses to real numbers, the commissioners and analysts were almost apologetic in their assertions:

> The item which yielded the highest number of "yes" responses is also the only double-barreled question in the survey. Unlike a simple query such as "Have you ever seen a UFO?", this question asks if two conditions have occurred simultaneously. One might therefore expect it to have a lower percentage of positive responses than the other, simpler items. Yet 18% of our respondents say they have wakened up paralyzed with a sense of a strange person or presence or something else in the room. This percentage, with a plus or minus 1.4 margin of error represents: 33,300,000 people.
>
> Those who have experienced a period of an hour or more in which they were apparently lost but did not remember why or where they had been: 24,050,000 people.
>
> Those that were actually flying through the air although they didn't know why or how: 18,500,000 people.

5 *Ibid.*, p. 25.

> Those who found puzzling scars on their bodies, with neither they nor anyone else remembering how they received them or where they got them: 14,800,000 people.[6]

The commissioners had included many of the paranormal queries because they knew from experience that such phenomena often accompany UFO and alien encounter activity, but they were not sure what the relationship was between UFOs and ghosts, for example.

Two percent of American adults answered affirmatively to four of the five indicator questions, resulting in a likelihood that no fewer than 3.7 million American adults have been abducted, according to the survey commissioners.[7] If true, this is a staggering result. Is it possible that 3.7 million American adults have been abducted by aliens in UFOs? The logistics involved in just picking up that many people, especially surreptitiously, are phenomenal.

Moreover, if the 2 percent statistic holds throughout the Earth's general adult population, then 100 million adults worldwide may have been likewise abducted in recent decades. That would indicate that in any given 24 hour period, thousands of adults are abducted. This represents a true logistical nightmare—our nighttime skies would be veritable freeways of alien activity—freeways in gridlock, because the numbers add up to nearly 10,000 abductions per night. We may chuckle at the prospect, but the fact remains that this is the frequency of activity being reported by sober-minded people—our relatives, friends, and neighbors. Any honest investigator must accept the fact that whether the phenomenon is extra-terrestrial or something else, it is extremely pervasive.

These numbers make us wonder if perhaps a growing segment of our population is being exposed to hallucinogens—or perhaps the drug abuse that began its now crescendoing ascent in the 1960s is beginning to show its cumulative ill-effects. The survey results cut across all demographic categories, however, although two groups, the PSAs and adults between the ages of 18 and 29, tended to report higher levels of UFO/paranormal experience. Although one might not be too surprised by this result at first glance, the analysts make some interesting points about it. Concerning the PSAs, the analysts speculate:

---

6 *Ibid.*, pp. 47-48.

7 *Ibid.*, p. 48.

> The positive correlation between social/political activism and indicator experiences came as a surprise to us. A priori, one might expect that more passive individuals would show more of the indicator experiences, yet the data shows the reverse. Could these experiences be causes of increased anxiety, and could this anxiety in turn be a motivator of social and political activities? It has also been theorized that this correlation does not necessarily mean that social/political actives are actually having more than the average number of unusual experiences. If respondents in this group are more confident and independent in their thinking, perhaps they are more willing than others to discuss unusual personal experiences. In this reading of the data the non-social/political actives may be under-reporting their unusual experiences.[8]

The analysts comment similarly about more youthful respondents:

> In virtually every case indicator experiences decline with age, especially after 44. This is puzzling, since older people would be expected to have accumulated more such experiences. One hypothesis is that indicator experiences are concentrated in the younger years—the time of greatest UFO abduction activity in an individual's life and are then forgotten with advancing age. We know from surveys of crime victims that memory of even major crimes tends to fade over time. It is also that the number of individuals undergoing UFO abductions is increasing year by year; since these events begin in childhood, the total number of abductees in each age group would be greater among the youngest groups.[9]

Most commentators tend to agree that the reason for higher numbers of younger respondents reporting these experiences is that the phenomenon itself has increased in its rate of activity in recent years. In fact, a quick comparison of the five indicator experiences by age is revealing.[10]

	18-29	30-44	45-59	60-+
Waking up paralyzed with sense of strange figure	22%	21%	17%	10%
Missing time	14%	13%	13%	10%
Feeling of actually flying	11%	13%	10%	8%
Balls of light in room	11%	9%	7%	5%
Puzzling scars	14%	7%	6%	5%

It is difficult to believe that such dramatic, and in many cases traumatic experiences could have been so easily forgotten with the mere passing of

8 *Ibid.*, p. 49.

9 *Ibid.*

10 *Ibid.*

years. This is especially true when the pertinent time periods are pre-geriatric years. The "puzzling scars" category is especially enlightening in view of the fact that in the youngest group 14 percent reported such anomalies, while in the very next age group, only half that number could account for the experience. If we were tempted to accept the theory that such memories fade with time, we would be hampered by the fact that there was no marked decline in the following age groups in that category—but only a gradual 1 percent decline thereafter.

Having reviewed the same data garnered by the experts, we ask, is there truly a confederation of scientists and other high level professionals who believe that millions of American adults have in recent decades been secretly abducted by aliens and taken aboard UFOs for examinations and related procedures, and now have no residual memory of the abductions? Do they accept as reality the indication that 100 million adults worldwide have been abducted? Apparently so.

A New Cultural Marvel

Library and bookstore shelves, as well as weekly television programs, are packed with stories telling and retelling of the hundreds of thousands of UFO sightings and close encounters that have been reported to date. Tales of UFO sightings, alien encounters, and human abductions recited are now by hard-core television journalists who only a few years ago were reporting on foreign wars from the trenches. Messages from our space brothers are copious, warning us of the impending destruction of our planet and steering us toward the cosmic gospel of the New Age. Tens of thousands of the world's citizens are being told by little grey men that they have been chosen to be intermediaries between the coming mother ships and the children of Earth. It is their mission to prepare humans first for imminent Earth changes that will destroy most human life, and then, for the merging of the species, to develop a new creature that will repopulate the planet.

Do these claims seem reasonable? Is it possible that any rational person would afford any degree of consideration to such declarations? The truth is that UFO books are the hottest selling nonfiction literature and the UFO programming on television is absorbed by many millions of American minds weekly. As with UFO sightings and abductions themselves, general interest in the phenomenon cuts across demographic boundaries. Inquiring minds want to know.

What is the net result of this incessant bombardment of the Western mind with messages from galactic messiahs? The answer to this question is the key to this work. As one can imagine, when the Harvard Medical School becomes so involved in promoting the reality of the phenomenon, in alliance with other ivy league schools and prestigious institutions, then we can assume that the general population has been fairly well indoctrinated, and is ready to believe.

With such heavy hitters now weighing in on the side of the contactees and abductees, there is little to doubt about the level of investigation into the alien encounter phenomenon. Only the most resolute debunkers still hold out hope that all UFO and alien encounter reports can be explained by naturally occurring phenomena, e.g., swamp gas, hoaxes, or mental disorders. The evidence is too prolific, too consistent, to be anything but conclusive that something is really occurring. The only question that remains is "what is happening to these millions of people?" Every now and then we uncover a hoaxer, a trickster who seeks the publicity or other rewards of being an abductee. For every hoaxer, however, there are tens of thousands of honest people who seek nothing, and who bear solemn testimonials to the truth of their terrifying encounters with the "beings."

Most people now know someone who has seen a UFO or has had an encounter with an unknown being. If you believe that you do not know such a person, just bring the subject up casually among your friends or relatives, and you will be surprised at the degree to which the phenomenon has penetrated your own life. There is no denying anymore that millions are being involved in something that does not fit well into our view of the world and its naturally occurring phenomena. Something is happening. What do the scientists, experts and other professionals conclude in their hundreds of books reporting on investigations into the phenomenon? Do they have a conclusion after so much effort and analysis? The collective conclusion of the origin of these beings that are penetrating our lives and culture has evolved over the past few decades. Those conclusions have altered our most fundamental beliefs about who we are.

Those who were bold enough in the early stages of the phenomenon to accept the possibility that encounters were really occurring, tended to entertain the extraterrestrial visitor theory nearly exclusively. After all, we have been culturally adapted to that possibility through our constant exposure to *Buck Rogers*, *Star Trek*, *Close Encounters of the Third Kind*, and *ET*, and the aliens themselves were telling us they hailed from far away planets and galaxies. Our media have provided us with an ongoing dialogue concerning the lifeforms in space and the complexities of encounters with them and exposure to them. It was not, therefore, difficult to give place for the extraterrestrial possibility.

Uncovering The Deception

As the "experts" have investigated the alien encounter phenomenon, however, this extraterrestrial theory seems to diminish in probability. The beings are now presenting themselves as something different than fleshy beings that have spawned from a foreign ocean, and flown here in nuts-and-bolts craft with superior propulsion systems.

This is the reason why these cutting-edge investigators asked the questions they did in the Roper Organization poll. Indeed, seeing or sensing traditionally spiritual phenomena and experiencing paranormal activity have become part-and-parcel of the UFO experience. Now, as demonstrated below in this work, investigators are concluding that although the beings present themselves as extraterrestrials with a mission, they are indeed more than just that—they are more than just foreign scientists or extraterrestrial anthropologists, here to study or forewarn. They appear to be presenting a false picture of themselves and their mission.

"Interdimensional" is the latest casting of the beings. The apology now in vogue is that they only appear "evil" and "sadistic" because they are fundamentally different from us and communicate dissimilarly from us. As we look deeply into the alien encounter phenomenon, we find that there is indeed "something wrong with this picture." The alien apologists' attempts to rehabilitate the acts of the "aliens" reminds one of the story of the Emperor's New Clothes. The Emperor was bilked into believing that his beautiful and costly new wardrobe was invisible. Everyone went along with the illusion, complimenting him on the fineness of his new apparel as he strolled naked through the streets. A child who eagerly waited with the crowds to see the Emperor's new clothes was stunned to see the naked Emperor and blurted out, "but the Emperor has no clothes." At this, the compliant crowd shuddered, wondering what would happen to this lad who refused to go along with the "new reality" of the Emperor's new clothes. The Emperor stopped and questioned the child, who responded in innocence that indeed, the Emperor had no clothes. With this challenge to the new reality created by the clothier, the Emperor suddenly accepted the true reality that he was naked before the throng, and the crowd began accepting the true reality of the circumstances and responded in kind. The Emperor was naked, and no amount of "new reality" could change that.

This book is an exercise in reality testing—or "new reality" testing. However, in looking more deeply into the alien encounter "presentation," we must ask ourselves certain questions, we must ask questions about ourselves and our universe, questions we normally prefer to relegate to other times, places, or people more expert than we in matters concerning "interdimensional" beings.

If these beings are truly not what they claim to be, we cannot only look at the evidence they present for answers, for their evidence must be designed to deceive. We must look elsewhere for answers because the "phenomenon" is achieving its desired effect: it is reeducating the Western world. After studying the encounter phenomenon in great depth, Dr. Mack, whom we would expect to be a clear thinker, sums up his conclusions and the reeducation he has received through dabbling in the phenomenon as follows:

> The experience of working with abductees has affected me profoundly. The intensity of the energies and emotions involved as the abductees relive their experiences is unlike anything I have encountered in other clinical work. The immediacy of presence, support, and understanding that is required has influenced the way I regard the psychotherapeutic task in general. Furthermore, I have come to see that the abduction phenomenon has important philosophical, spiritual, and social implications. Above all, more than any other research I have undertaken, this work has led me to challenge the prevailing worldview or consensus reality which I had grown up believing and had always applied in my clinical/scientific endeavors. [p. 3]
>
> There can be little place, especially within the Judeo-Christian tradition, for a variety of small but powerful homely beings who administer an odd mixture of trauma and transcendence without apparent regard for any established religious hierarchy or doctrine. It is one thing to acknowledge that "spirit" resides in the universe and "we are not alone." It is quite another for "spirit" to show up in such odd and threatening form, created partially in our own image. At best, this would seem puzzling and difficult to integrate. At worst, to the polarizing perception of Christian dualism these dark-eyed beings must seem to be the playmates of the Devil (Downing 1990). Eastern religious traditions, such as Tibetan Buddhism, which have always recognized a vast range of

> spirit entities in the cosmos, seem to have less difficulty accepting the actuality of the UFO abduction phenomenon than do the more dualistic monotheisms, which offer powerful resistance to acceptance. [p. 412]

Western philosophy does, in fact, give place for these encounters and their related phenomena. There is positively something wrong with the picture being presented by the alien encounter phenomenon—indeed, the Emperor has no clothes. Although the "aliens" keep attempting to mislead us in our beliefs about them and their purpose, serious questions about their "presentation" continue to arise in the minds of those so encountered. Dr. Mack echoes the private confidences of many abductees in revealing, "Quite a few abductees have spoken to me of their sense that at least some of their experiences are not occurring within the physical space/time dimensions of the universe as we comprehend it. They speak of aliens breaking through from another dimension, through a 'slit' or 'crack' in some sort of barrier, entering our world from 'beyond the veil.'" (p. 404) The "Near Death Experience" literature is very similar in its descriptions of encounters with beings from "beyond the veil," yet we are culturally reticent to publicly compare the two because of the spiritual implications.

Famed author and abductee, Whitley Strieber, meets some these questions head on in his book *Transformation: The Breakthrough.*[11] He speaks for many who have endured these encounters as he describes his relationship with the beings.

> Increasingly I felt as if I were entering a struggle that might be even more than life-or-death. It might be a struggle for my soul, my essence, or whatever part of me might have reference to the eternal.
>
> There are worse things than death, I suspected. And I was beginning to get the distinct impression one of them had taken an interest in me.
>
> So far the word *demon* had never been spoken among the scientists and doctors who were working with me. And why should it have been? We were beyond such things. We were a group of atheists and agnostics, far too sophisticated to be concerned with such archaic ideas of demons and angels. [pp. 44-45]

Are we so sophisticated that we are unable to speak of interdimensional beings, demons, or even angels in mixed company? Are these subjects reserved only for crackpots and day time talk shows? Does the fact that

---

11 Beechtree Books/William Morrow, New York (1988).

millions of American adults are encountering such beings give us reason to pause and consider their testimonials, or does it drive us further and faster into the accelerating pace of modern life in our collective effort to avoid the truly "big" questions? "Alien" and similar "being" encounters are plaguing modern civilization at epidemic levels. Yet, we choose to ignore these prolific "attacks," officially anyway, because their very existence reminds us that our roots are closer to the desert prophets and aboriginal shamans than we care to believe.

The alien encounter phenomenon dances along the outer boundaries of acceptable beliefs, one foot within and one without, goading us with its threats, mocking us with its power and display, and changing us with its message. We believe because of the testimonials, but we reject the phenomenon because of its absurdity. We waiver, we tremble, we embrace —but we do not openly discuss it, because at its deepest levels, we see that the Emperor has no clothes, and that we are facing a modern surge of something that our forefathers feared and avoided throughout recorded history.

The Approach Taken In This Book

No other planets have yet been discovered by astronomers outside of our own solar system, with a possible exception of one in early 1992. Yet many scientists postulate that there exist numerous planets circling myriad stars in our universe. Of these planets, some estimate that a low percentage (still, thousands of planets) are capable of sustaining life. Although it would be statistically nearly impossible for life to exist on such planets from a purely evolutionary viewpoint,[12] Biblically, we believe that God creates life, and places it where it suits Him. Armed with this knowledge, it would be easy for us to say "yes," we are likely being visited by our brothers from one or more of the distant planets on which God has possibly placed life made in His image. There are, however, certain patterns to the forms of visitation, the forms of communication, and the content of the communication from the bulk of these UFO and alien encounters that give us great pause in saying "welcome, brothers!"

---

12 *Nature,* "UFO, APPENDIX," by Dr. Frank B. Salisbury, October 1969, p. 2.

Are UFOs real objects—does their presence represent a physical phenomenon? Most scientists found it too difficult initially to accept the existence of UFOs. Many, however, came to find the evidence too compelling to ignore, and began accepting the possibility of extraterrestrial life and its visitation to Earth. Many of these, with no particular philosophical training or background, for the reasons discussed herein, are beginning to rethink the "extraterrestrial visitors" assumption. Now, having accepted the possibility of extraterrestrial life, they, like laymen, are finding great difficulty in ascertaining whether or not UFOs are manifestations of extraterrestrial visitation, or something else.

To unravel the tangle of "UFOlogy" (the study and body of UFO reports) and how it relates to modern Western life, this book surveys the UFO literature[13] and looks closely at those who are "victims" of close encounters with UFOs and their occupants. In so doing, we examine the history of UFO phenomena, in their various forms and manifestations, including "alien" manifestations at all levels. We also analyze the "message" and other information we are receiving from UFOs and related sources, and weigh in the balance the nature, origin, and character of these, our "space brothers."

Reaction to the material contained in this work is always potent. Many readers thrill to every page, feeling strongly the spirit of the warning message contained herein. Others, however, have initially been shaken, frightened by the implications of the research—perhaps finding it difficult to sleep alone for a few nights.

There is no intention of heartlessly terrifying the reader by sharing these things. The information and analysis are offered as education. The UFO phenomenon is real and widespread and will have far-reaching effects in coming years, although its true nature and origin are yet unproved. An attempt is made herein to fully document the cases and testimonials that provide a database, and offer the conclusions of the world-class experts that have investigated them. The reader is invited to analyze the data in a thoughtful manner. Whether or not your conclusions are the same as those offered herein, by reading these materials you will at least have received an education regarding a phenomenon that will take on greater prominence and significance in the months and years in our very near future.

---

13 The initial survey is made without analysis or criticism, for the purpose of establishing the popular conception of what is occurring in the UFO/alien encounter phenomenon. Analysis of the reports and the possible geneses of the phenomena follow.

2
UFOs In Antiquity

UFO Reports Throughout The Ancient World

From the time man learned to write down on parchment or clay tablets the important details of his day, brief accounts of mysterious flying vessels have been inserted between other affairs of state. Many believe that various accounts in the Bible have such origins. In ancient China, a legend of humanoid visitors in "flying carts" with gilded wings found its own genesis, and the tale has been passed on for centuries. In Rome, "flying shields," sometimes accompanied by spurting fire, were widely reported. Written in the Sanskrit *Drona Parva* text are accounts of a superhuman race conducting aerial dogfights in flying vessels called "vimana." One translation renders specifics of an air battle thus: a "blazing missile possessed of the radiance of smokeless fire was discharged."

Two of the most famous remote reports of flying vessels are those that occurred in the skies over Nuremburg, Germany, and Basel, Switzerland, in 1561 and 1566, respectively. During these spectaculars, thousands of local citizens were treated to a display of several large disks, spheres, and tubes which appeared in the skies, and "danced" or weaved themselves about in something described as an aerial ballet. When finished, these objects suddenly resolved themselves into "fiery red" spheres, and disappeared. Numerous historical reports like these have come to light in recent years.

Scriptural Accounts?

There exist many books touting mankind's "alien" origins and claiming that the Bible is laden with tales of UFO visitations to Earth. These writers cite scriptures, generally from the Bible's Old Testament, that refer to flying chariots and other objects as evidence of divine dependence on mechanical vehicles for mobility not only between planets, but also on the Earth's surface. Although this work does not endorse these theories, hereafter are a few of the accounts generally proffered in behalf of the hypothesis.

Chariots Of Fire

The assumption of the prophet Elijah is frequently cited and held to be especially descriptive of an ancient flying chariot. "And it came to pass, as they still went on, and talked, that, behold, *there appeared* a chariot of fire, and horses of fire, and parted them both asunder; and Elijah went up by a whirlwind into heaven. And Elisha saw *it*, and he cried, My father, my father, the chariot of Israel, and the horsemen thereof. . . ." (2 Kings 2:11-12.) Biblical annotators cross-reference this passage with Isaiah 66:15, which says: "For, behold, the Lord will come with fire, and with his chariots like a whirlwind, to render his anger with fury, and his rebuke with flames of fire."

Although we can somewhat easily interpret God's elemental chariots as being merely descriptive of His comings and goings as being with power and celebration, the chariot of fire that took Elijah into heaven may have a more functional character. A look at the sight beheld and described by Elisha includes "horses of fire" complete with "horsemen." UFO theorists say that these are mere descriptions of mechanical apparatus that propelled the craft, but it could well be that Elisha was shown a vision that accompanied Elijah's assumption, for the purpose of imbuing the occasion with power and authority which has been the practical result. There are those who are always quick to explain away scriptural, or even miraculous reports by "interpreting" them away, saying that the report is phrased in the terms and understanding of the time. Even if true, however, this is not a license to interpret every flying object or miracle as having a UFO origin which has become the practice of these "angels=aliens" protagonists. "Horses" may just mean "horses."

The Camp Of Israel Led By A UFO

Another proffered evidence of the Biblical UFO phenomenon is the pillar that accompanied Israel on its exodus from Egypt, across Sinai, to the Promised Land.

> And the Lord went before them by day in a pillar of a cloud, to lead them the way; and by night in a pillar of fire, to give them light; to go by day and night: He took not

> away the pillar of the cloud by day, nor the pillar of fire by night, *from* before the people. [Exodus 13:21-22]
>
> And the angel of God, which went before the camp of Israel, removed and went behind them; and the pillar of the cloud went from before their face, and stood behind them:
>
> And it came between the camp of the Egyptians and the camp of Israel; and it was a cloud and darkness *to them,* but it gave light by night *to these*: so that the one came not near the other all the night.
>
> And Moses stretched out his hand over the sea; and the Lord caused the sea to *go back* by a strong east wind all that night, and made the sea dry *land,* and the waters were divided.
>
> And the children of Israel went into the midst of the sea upon the dry ground: and the waters were a wall unto them on their right hand, and on their left.
>
> And the Egyptians pursued, and went in after them to the midst of the sea, even all Pharaoh's horses, his chariots, and his horsemen.
>
> And it came to pass, that in the morning watch the Lord looked unto the host of the Egyptians through the pillar of fire and of the cloud, and troubled the host of the Egyptians,
>
> And took off their chariot wheels, that they drave them heavily: so that the Egyptians said, Let us flee from the face of Israel; for the Lord fighteth for them against the Egyptians. [Exodus 14:19-25]

These verses inform us of some interesting characteristics of this pillar. First, we are told that the *Lord* went before Israel in the pillar. We can understand this to mean that He led them by the means of the pillar. However, we are next told that "the angel of God, which went before the camp of Israel, removed and went behind them." This gives us a more focussed view of the nature of the pillar, that it was an angelic presence. UFO theorists conjecture that the pillar was possibly UFO exhaust, as evidenced by its dark cloudiness at times, and light at others. Therefore, they conclude that the "angel of God" was in a craft or device that exhausted a misty substance that could be lit up or left dark. Protagonists further claim that the "east wind" that blew and parted the sea and dried its floor in a single night was an antigravity field generated by the craft. This antigravity theory, they say, is supported by the fact that the Lord was able to remotely take "off their chariot wheels." And finally, because the Lord "looked unto the host of the Egyptians through the pillar of fire and of the cloud" as he troubled them with these difficulties, the pillar was more of a device than a presence.

The Wheels Of Ezekiel

The most vivid of all Biblical descriptions of flying "vehicles" is that of the prophet Ezekiel. With a little imagination, we can appreciate why Biblical UFO adherents believe that Ezekiel saw a genuine flying craft.

And I looked, and, behold, a whirlwind came out of the north, a great cloud, and a fire infolding itself, and a brightness was about it, and out of the midst thereof as the colour of amber, out of the midst of the fire.

Also out of the midst thereof *came* the likeness of four living creatures. And *this* was their appearance; they had the likeness of a man.

And every one had four faces, and every one had four wings.

And their feet were straight feet; and the sole of their feet was like the sole of a calf's foot: and they sparkled like the colour of burnished brass.

And *they had* the hands of a man under their wings on their four sides; and they four had their faces and their wings.

Their wings *were* joined one to another; they turned not when they went; they went every one straight forward. As for the likeness of their faces, they four had the face of a man, and the face of a lion, on the right side: and they four had the face of an ox on the left side; they four also had the face of an eagle.

Thus *were* their faces: and their wings *were* stretched upward; two *wings* of every one *were* joined one to another, and two covered their bodies.

And they went every one straight forward: whither the spirit was to go, they went; *and* they turned not when they went.

As for the likeness of the living creatures, their appearance *was* like burning coals of fire, *and* like the appearance of lamps: it went up and down among the living creatures; and the fire was bright, and out of the fire went forth lightning.

And the living creatures ran and returned as the appearance of a flash of lightning.

Now as I beheld the living creatures, behold one wheel upon the earth by the living creatures, with his four faces.

The appearance of the wheels and their work *was* like unto the colour of a beryl: and they four had one likeness: and their appearance and their work was as it were a wheel in the middle of a wheel.

When they went, they went upon their four sides: *and* they turned not when they went.

As for their rings, they were so high that they were dreadful; and their *rings* were full of eyes round about them four.

And when the living creatures went, the wheels went by them: and when the living creatures were lifted up from the earth, the wheels were lifted up.

Whithersoever the spirit was to go, they went, thither *was their spirit* was to go; and the wheels were lifted up over against them: for the spirit of the living creature *was* in the wheels.

When those went, *these* went; and when those stood, *these* stood; and when those were lifted up from the earth, the wheels were lifted up over against them: for the spirit of the living creature *was* in the wheels.

And the likeness of the firmament upon the heads of the living creature was as the colour of the terrible crystal, stretched forth over their heads above.

And under the firmament *were* their wings straight, the one toward the other: every one had two, which covered on this side, and every one had two, which covered on that side, their bodies.

And when they went, I heard the noise of their wings, like the noise of great waters, as the voice of the Almighty, the voice of speech, as the noise of an host: when they stood, they let down their wings.

And there was a voice from the firmament that *was* over their heads, when they stood, *and* had let down their wings.

And above the firmament that was over their heads *was* the likeness of a throne, as the appearance of a sapphire stone: and upon the likeness of the throne was the likeness as the appearance of a man above upon it.

And I saw as the colour of amber, as the appearance of fire round about within it, from the appearance of his loins even upward, and from the appearance of his loins even downward, I saw as it were the appearance of fire, and it had brightness round about.

As the appearance of the bow that is in the cloud in the day of rain, so was the appearance of the brightness round about. This *was* the appearance of the likeness of the glory of the Lord. And when I saw *it,* I fell upon my face, and I heard a voice of one that spake. [Ezekiel 1:4-28]

Ezekiel continues his narrative in Chapter 10, describing the "object's" takeoff.

THEN I looked, and, behold, in the firmament that was above the head of the cherubims there appeared over them as it were a sapphire stone, as the appearance of the likeness of a throne.

And he spake unto the man clothed with linen, and said, Go in between the wheels, *even* under the cherub, and fill thine hand with coals of fire from between the cherubims, and scatter *them* over the city. And he went in in my sight.

Now the cherubims stood on the right side of the house, when the man went in; and the cloud filled the inner court.

Then the glory of the Lord went up from the cherub, *and stood* over the threshold of the house; and the house was filled with the cloud, and the court was full of the brightness of the Lord's glory.

And the sound of the cherubims' wings was heard *even* to the outer court, as the voice of the Almighty God when he speaketh.

And it came to pass, *that* when he had commanded the man clothed with linen, saying, Take fire from between the wheels, from between the cherubims; then he went in, and stood beside the wheels.

And *one* cherub stretched forth his hand from between the cherubims unto the fire that was between the cherubims, and took *thereof,* and put *it* into the hands of *him that* was clothed in white linen: who took *it,* and went out.

And there appeared in the cherubims the form of a man's hand under the wings.

And when I looked, behold the four wheels by the cherubims, one wheel by one cherub, and another wheel by another cherub: and the appearance of the wheels was as the colour of a beryl stone.

And *as for* their appearances, they four had one likeness, as if a wheel had been in the midst of a wheel.

When they went, they went upon their four sides; they turned not as they went, but to the place whither the head looked they followed it; they turned not as they went.

And their whole body, and their backs, and their hands, and their wings, and the wheels, *were* full of eyes round about, *even* the wheels that they four had.

As for the wheels, it was cried unto them in my hearing, O wheel.

And every one had four faces: the first face *was* the face of a cherub, and the second face *was* the face of a man, and the third face of a lion, and the fourth face of an eagle.

And the cherubims were lifted up. This is the living creature that I saw by the river of Chebar.

And when the cherubims went, the wheels went by them: and when the cherubims lifted up their wings to mount up from the earth, the same wheels also turned not from beside them.

When they stood, *these* stood; and when they were lifted up, *these* lifted up themselves *also:* for the spirit of the living creature *was* in them.

Then the glory of the Lord departed from off the threshold of the house, and stood over the cherubims.

And the cherubims lifted up their wings, and mounted up from the earth in my sight: when they went out, the wheels also *were* beside them, and *every one* stood at the door of the east gate of the Lord's house; and the glory of the God of Israel was over them above.

This is the living creature that I saw under the God of Israel by the river of Chebar; and I knew that they *were* the cherubims.

Every one had four faces apiece, and every one four wings; and the likeness of the hands of a man *was* under their wings.

And the likeness of their faces *was* the same faces which I saw by the river Chebar, their appearances and themselves: they went every one straight forward. [Ezekiel 10:1-22]

Many UFO writers attempt to prove a "UFO/history/religion" connection from Ezekiel's vision. They further generally offer an artist's conception of the "craft" described in great detail in the scripture. NASA *Saturn V* engineer, Josef F. Blumrich, attempted in 1968 to debunk the theory that "Ezekiel's Wheel" was a flying craft. He published a book in 1973 entitled *The Spaceships of Ezekiel,* in which he described how his debunking effort became a successful design venture. He writes, "Seldom has a total defeat been so rewarding, so fascinating, and so delightful!" His design is simple, resembling a domed upside-down loudspeaker with four vertical landing gear adorned with propellers. Although this type of undertaking might be enjoyable to an engineer, it is incongruous that beings that are capable of intergalactic flight would fly around in Earth's atmosphere in propeller-driven craft.

As a footnote to Ezekiel's experience, Ezekiel reports that he encountered an oddly colored "man" in connection with this "craft." "And he brought me thither, and, behold, *there was* a man, whose appearance was like the appearance of brass, with a line of flax in his hand, and a measuring reed; and he stood in the gate." (Ezekiel 40:3; cross-referenced from Ezekiel 1:7 (color of craft's feet.) The "flax" or "measuring reed" in the fellow's hand is a point of interest to UFO investigators. UFO occupants are sometimes said to carry silver rods in their hands, or even flax or reeds.

Von Daniken's "God And Angels Were Aliens" Theory

The premier champion of this "angels=aliens" theory is the ever prolific Erich von Daniken, best known for his book, and the motion picture based thereon, *Chariots of the Gods.* In his book, *Signs of the Gods,* von Daniken, as in his other dozen or so books on the subject, pursues his extraterrestrial deity through any scripture or other written record that makes any mention of flight, fire, light, cloud, transportation, vision, visitation, markings on the ground, art, religion, legend, or miraculous occurrence. His critics sometimes assert that his quantum leaps in logic and unfounded philosophical skyscrapers demonstrate that he is one of the least intellectual writers of our time. However, in his untiring efforts to justify his position on extraterrestrial sources of our religious history, he occasionally flushes out some interesting tidbits of research. However, one must always separately authenticate von Daniken's research claims, because he is well known for "overstating" the facts.

Von Daniken cites *The Book of Enoch,* wherein God chastises the angels: "Why have you done like the children of earth and begot giant sons?" In

Chapter 18 of *The Book of the Secrets of Enoch,*[1] Enoch speaks of the "Grigori," a race of "soldiers" with human appearance held at bay in the fifth heaven, being of gigantic size, whose faces were withered and melancholy and whose mouths were silenced. Enoch is told that three of these followers of Satanail, the one who rebelled, broke through to the Earth and "saw the daughters of men how good they are, and took to themselves wives, and befouled the earth with their deeds, who in all times of their age made lawlessness and mixing, and giants are born and marvelous big men and great enmity." (Verses 1-6)

Von Daniken further cites in support of his "angels=aliens impregnated Earth women" theory, Chapter 100 of *Kebra Nagast,* which he claims to be an ancient Ethiopian writing, which purportedly says: "But every daughter of Cain with whom the angels had consorted became pregnant, but could not give birth and died. And of the fruits of their wombs some died, and others came forth; they split their mother's womb and came out of the navel. When they were older and grew up, they became giants . . . ." (*Signs of the Gods,* p. 117) Of course, von Daniken's giants are the very same that appear throughout mythology as cyclops, etc., and must, therefore, be the very beings responsible for the famous "ruts" and megalithic temples of Malta. This is how von Daniken reasons, anyway.

The concept of giants being born to normal human women impregnated by "watchers" or "sons of Heaven" is quite controversial among Christian scholars and UFOlogists alike. Our first indication of the enigma is found in Genesis 6:4, which reads: "There were giants in the earth in those days; and also after that, when the sons of God came in unto the daughters of men, and they bare *children* to them, the same *became* mighty men which *were* of old, men of renown." Noah's father, Lameck, was worried that Noah may have been the offspring of one of these enigmatic "watchers" or "sons of God," as reported in the Dead Sea Scrolls. Lameck was not alone in these assumptions. The subject pops up in scholarly literature and dissertation throughout recorded history, ancient and modern. Although not canonical, the passage is interesting, and the reader is free to determine the value of the information:

---

1 *The Apocrypha and Pseudopigrapha of the Old Testament in English,* R. H. Charles, D. Litt., D.D., Ed., (Oxford at the Clarendon Press: 1913), Vol. II.

> An extremely unusual son was born to Lameck. The child's body was white as snow, with parts as red as a rose. His long hair was white as wool, and his eyes were piercing and brilliant. He was able to talk immediately and, according to Lameck, apparently conversed with the Lord. Lameck, concerned and disturbed, wondered if the boy were his own or possibly had been conceived by one of the "watchers" or "sons of heaven." He discussed the matter with his wife, BatEnosh, who swore that the boy was Lameck's. Lameck took his problem to his father, Methuselah, who in turn sought counsel from his father, Enoch, who previously had been taken (translated) into heaven. Enoch told Methuselah to assure Lameck that his son had been sent from God to do a great work on the earth and that his name should be called Noah.[2]

Some Christian scholars believe that these "sons of God" were a certain class of angels set on earth as watchers, even protectors, who became renegades and began indulging in inappropriate earthly activities.[3] Some UFOlogists believe that these sons of heaven were large humans from another planet, who, during a mission to Earth, impregnated many earth women resulting in the birth of giant human children.

In support of the "angels=aliens" theory, proponents point us to the Old Testament Hebrew word for giants used in Genesis, "nephil" or "naphal," which they say literally means "the fallen ones," or, "those who are fallen down." They infer from this strained definition that these are beings who came from the sky. As intriguing as this sounds, a look at a Hebrew and Chaldee Dictionary[4] reveals that nephil merely means a feller, or lumberjack, and naphal means one who has failed or has been overcome. Although there were and are giants on the Earth, they most likely have not resulted from unions of aliens or angels with human women.

Enoch's Escort

In the beginning of *The Book of the Secrets of Enoch,* the 365-year-old prophet recounts an unusual event that gave rise to one of the most

---

2 *Christ's Eternal Gospel,* p. 155, rendering parallel translations from *The book of Lameck, The Book of Noah,* and *The book of Enoch* from the scroll "A Genesis Apocryphon," the Dead Sea Scrolls.

3 This position is somewhat strengthened by the verses of Genesis immediately following the statement, which reflect God's determination to destroy the Earth's inhabitants because of their great wickedness.

4 *See e.g., Strong's Exhaustive Concordance of the Bible, Dictionary of the Hebrew Bible,* p. 79, words nos. 5303-5307.

fantastic prophetic encounters in history. Again, the Enochian texts are not canonized by any Judeo-Christian sect, but the narration is instructive.

> And when I was asleep, great distress came up into my heart, and I was weeping with my eyes in sleep, and I could not understand what this distress was, or what would happen to me. And there appeared to me two men, exceeding big, so that I never saw such on earth; their faces were shining like the sun, their eyes too were like a burning light, and from their lips was fire coming forth with clothing and singing of various kinds in appearance purple, their wings were brighter than gold, their hands whiter than snow. They were standing at the head of my couch and began to call me by my name. And I arose from my sleep and saw clearly those two men standing in front of me. And I saluted them and was seized with fear and the appearance of my face was changed from terror, and those men said to me: 'Have courage, Enoch, do not fear; the eternal God sent us to thee, and lo! thou shalt to-day ascend with us into heaven, and thou shalt tell thy sons and all thy household all that they shall do without thee on earth in thy house, and let no one seek thee till the Lord return thee to them. . . .
>
> It came to pass, when Enoch had told his sons, that the angels took him on to their wings and bore him up on to the first heaven and placed him on the clouds. And there I looked, and again I looked higher, and saw the ether, and they placed me on the first heaven and showed me a very great Sea, greater than the earthly sea.
>
> They brought before my face the elders and rulers of the stellar orders, and showed me two hundred angels, who rule the stars and their services to the heavens, and fly with their wings and come round all those who sail. [Chapters 1, 3 and 4]

Enoch relates in later chapters how he was shown the workings of the planets and stars and the myriad angels assigned to oversee those functions. The most pertinent aspect of these accounts is the description that Enoch gives us concerning the appearance of the angels and their mode of transportation. His insistence that the angels possessed "wings" which enabled them to fly from one heaven to another is fascinating. Could this indicate that they used something with wings for transportation as some would have us believe? Functionally, wings are not necessary for flight—not interstellar flight anyway—not for God, angels, or UFOs.

UFOs In Egyptian Religious Texts

Although not referring to UFOs, in drawing parallels between the ancient Gospel and other religions of antiquity, a professor of ancient languages and cultures, Hugh Nibley, makes reference to accounts of interplanetary visitations by beings from other worlds who commute for purposes related to the spiritual welfare of humans.

Professor Nibley adds to this a colorful Ancient Egyptian description of ancient spiritual space travel: "It is in the solar ship that the initiate joins his father on the horizon. One steps into the Sun-ship just as it reaches that place where the sky touches the earth at the horizon as the water meets the land at the sacred wharf of the pyramid or temple. It is a ship that carries one in a state of effortless suspension through the void between the worlds."

Remembering that all of this describes the transporting of the dead king's soul to his eternal resting place, much like the Christian heaven, the resemblance to reports by abductees of the "state of effortless suspension through the void between the worlds" is remarkable. Perhaps they have a common origin, as we discuss below.

One last interesting aspect discussed by Professor Nibley is the purported means of propulsion of these great Solar Barks. He relates that, "As the ship moves on, we remember that after all it is the Solar-bark, a Skyship, 'Moving in Light' (Thausing, Gr. Tb., p. 9)." In UFO literature we have many accounts of UFO occupants describing the pro-pulsion system of their ship to their guest. One such description is that of a conservative Christian man, Udo Wartena, who's account is given in Chapter 5. His hosts explained the mechanics of their craft to him, and then revealed to him that they focused on a distant star and used its energy to draw them through space at speeds greater than the speed of light. He then adds, "My host specifically mentioned 'skipping upon the light waves.'" The correlation between "skipping upon the light waves" and "moving in light" seems substantial.

"Alien" Encounters Throughout History?

Famed UFO investigator and author Jacques Vallee has documented many similarities and parallels between the medieval "fairy faith" and modern accounts of UFO sightings, encounters, and abductions. The peasantry were so overwhelmed by their encounters with small humanoids who whisked them away, generally into underground complexes, and performed physical examinations as well as reproductive experimentations, that they revered these "fairies" as a supernatural force to be reckoned with. One village documented how one of its prominent midwives was kindly abducted and taken to an underground area where a humanoid giving birth needed assistance. When the ordeal was over, the midwife was thanked, and returned to her village to tell the tale. Accounts like these are disturbingly similar to modern abduction accounts. Further abduction materials, traditional and alien, are discussed at length in the following chapters.

Having reviewed a cross-section of historic sightings and encounters, more questions about the phenomena arise than are answered. We do observe certain parallels, however, that begin to take form. We will see these as we survey the phenomena in the remaining chapters. As to the specific question about whether God and angels require the protective hulls of metallic spacecraft for their comings and goings throughout the cosmos, perhaps the most lucid treatise on the subject was delivered by the immortal Captain James T. Kirk in the motion picture *Star Trek V,* wherein he tested the new reality of the scene and queried, "Excuse me—excuse me but, what does God need with a Federation star ship?"

3
Modern UFO Sightings

UFO Reports In Recent History

The current era of UFO sightings actually began in the late nineteenth century—around the period of great airships. With popularized reports of high-tech flying machines hitting the public press, stories of flying machine sightings proliferated into an American avocation. Although many such accounts were accurate reports of prototype gas-filled balloon craft flying overhead, making mysterious noises and sporting eerie, colored lights, others were preposterous, and obvious fabrications intended to elicit attention to the reporters. First, newspaper reporters were known to fabricate such tales to boost circulation, then telegraph operators did it for the pure pleasure of telling the tales.

With the advent of giant airships developed by industrialists bent on assisting with skirmishes in Cuba and elsewhere around the world, tales of peculiar night encounters with futuristic voyagers abounded. Although it is true that a handful of experimental airships nocturnally[1] crisscrossed the United States at speeds of twenty to thirty miles per hour, tales of ultra-modern craft moving at 200 miles per hour (or much faster) surfaced sporadically. These tales gave rise to accounts of people witnessing hieroglyph-speckled crafts piloted by superhuman beings clad in space-type uniforms. Many reports, made by credible citizens of the period, were very similar to UFO accounts of today rapid movements and departures and emanating light beams that probed and lifted.

UFO Expectations

As of 1900, physical descriptions of these mysterious, advanced craft were limited to airships of the period—the dirigible kind; no flying saucers and no "aliens," yet. This "airship" portion of UFO history illustrates a dis-

---

1 Experimental flights of these craft were conducted at night because hot air balloons fly better in cold weather, and supposedly, to protect pending patents.

turbingly consistent phenomenon that has perplexed researchers for decades. From the late nineteenth century through the present, the UFO phenomenon, although consistent in other respects, has manifested itself in the technological and cultural (including sciencefictional) trappings and expectations of the era.

As noted and discussed by Dr. Salisbury in *The Utah UFO Display,*[2] UFOs appear to people of different cultures, beliefs, and experiences, in manners seemingly diverse than to those of dissimilar backgrounds. For example, the encounters at Fatima, Portugal, discussed later in this work, were "packaged" specifically for rural Portuguese Catholics of the early twentieth century. Other encounters, in South America for example, appear to be tailored to Catholics of that region. There is a slight difference in belief, and therefore, cultural expectation. The astute reader may say, "But the Fatima and other 'Madonna' sightings appear more like spiritual manifestations of some kind rather than a UFO experience." This observation is one that should be kept in mind as all UFO sightings are analyzed. UFO author Brad Steiger has observed, "A historical survey reveals that reports of strange objects in the skies are laced through documents of the ancient and recent past. Interestingly, the records seem to indicate that UFOs have adapted themselves to the cultural milieu and the technological capacities of the observers."[3] As stated frequently herein, many UFO researchers are turning from their "nuts and bolts extraterrestrial flying machine" explanations of UFO origins to more metaphysical interpretations. The line between "nuts and bolts" and "spiritual" becomes quite hazy as the overall UFO occurrence is examined.

The "tailoring" of these divergent manifestations becomes somewhat apparent in the following chapters. The differences are not limited to those of perception, either. The craft, the humanoids, and the "UFO message" and its means of delivery are manifested in a diverse, yet somewhat predictable fashion. This is why Dr. Salisbury employs the word "display" in his title to illustrate the "theatrical" aspects of the encounters. As we begin our survey of modern UFO accounts, you are invited to note this pattern, as well as others that you detect, and to discern for yourself the paradigm of the UFO encounter.

---

2 Frank B. Salisbury, Ph.D., *The Utah UFO Display: A Biologists Report,* Devin-Adair, Connecticut, 1974.

3 Brad Steiger, *The UFO Abductors,* Berkley Publishing Corporation, New York, 1988, p. 212.

UFOs In The Twentieth Century

The first half of the twentieth century saw the rapid escalation of reports of UFOs, but again, the accounts were initially limited to airships, and later, airplane-like craft and rockets. However, by now, popular literature had become overrun with Jules Verne and H. G. Wells "wannabees," and science fiction paperbacks became an American staple. Popular writers and radio programs weaved tales of air and space travel that captured the imagination of most of America, and then the world. Stories of close encounters with aliens from Mars, Venus, Jupiter, and even the Moon became common. Dime novel and mystery magazine publishers did all they could to promote their science fiction bonanza by creating eye-witness accounts of exotic interplanetary rockets manned by inhuman aliens; bug-eyed, gill-breathing monsters, whose designs were generally imperialistic.

As the skies began to fill with shiny airplanes, test rockets, and other experimental craft, and as the minds of the public became habituated to the idea of interplanetary space travel, reports of weird flying machines abounded around the world. These reports were often the creations of unscrupulous reporters and publishers wishing to boost careers and increase circulation. By the time H. G. Wells' 1898 book *War of the Worlds* hit the radio airwaves, the public was willing to believe. Other sighting narratives made by average citizens of the time were somewhat less sensational and often mistaken, yet reported by believable townsmen who lent credibility to the UFO "industry."

Public UFO Reports By Credible Witnesses

During World War II allied, as well as enemy—pilots and crewmen, reported seeing small globes following them as they flew in formation. These "foofighters" (also known as "Kraut Balls") as the allied airmen came to call them, were thought by both sides to be the secret weapon of the other. Fliers complained of being followed by one to ten of these red, orange, or clear spheres, believing them to be the result of superior German technology. Allied pilots soon learned, however, that the spheres were harmless. Former B-17 bomber pilot Charles Odom told the Houston *Post* that the foofighters "looked like crystal balls, clear, about the size of basketballs" and that they would advance to within about 100 yards of the flight formation and "would seem to become magnetized to our formation and fly alongside. . . . After awhile, they would peel off like a plane and leave."[4] These observations were so wide-spread that most World War II fliers reportedly experienced them. Subsequent to World War II, many pilots continue to report the presence of these luminous, enigmatic spheres. This is especially true in aerial combat situations, such as in Korea and Vietnam.

The Modern Era Of UFOs

The modern era of UFOs was ushered in on June 24, 1947, when Kenneth Arnold, a respectable Boise, Idaho pilot, observed what he described as "nine peculiar aircraft" flying in formation and maneuvering at approximately 1,700 mph,—a speed unattainable in 1947. Arnold was a sober professional and experienced aircraft observer. His mission that day was to locate a downed C-46 Marine transport plane. He was watching very carefully when the nine metallic discs came into view. Arnold rejected the publicity that came his way as a result of what he saw on that clear summer day. He had assumed that the unidentifiable aircraft belonged to the U.S. Military, and fully expected a reasonable explanation of why disc-shaped craft without tails were flying at phenomenal speeds near Mt. Rainier, Washington, just as soon as he contacted the General Manager of Central Aircraft at the Yakima Airport. The General Manager was slightly dubious of Arnold's report, however, which set Arnold on a path leading to Military Intelligence to learn if unmanned guided missiles were being fired in the region. There was none.

In the next hours and days, hundreds of reports flooded the air traffic airwaves and the media that others had seen clusters of Arnold's flying "saucers." This proliferation of flying saucer reports caused Arnold to later write:

> From then on, if I was to go by the number of reports that came in of other sightings, of which I kept a close track, I thought it wouldn't be long before there would be one of these things in every garage. In order to stop what I thought was a lot of foolishness, and since I couldn't get any work done, I went out to the airport, cranked up my plane, and flew home to Boise.

Arnold attempted a few times to investigate what he considered to be credi-

---

4 Raymond E. Fowler, *The Watchers: The Secret Design Behind UFO Abductions,* Bantam Books, 1990, p. 70.

ble UFO sightings. However, when he learned that sensationalist publishers were using his name to promote their books and magazines, he gave it up.

Arnold's sighting and report of June 24, 1947, remains one of the most credible UFO sighting reports to date.[5] Even Military Intelligence accepted Arnold's story as completely true, without concluding the origin of the discs. Even then, however, UFO debunkers were on the scene attempting to prove that Arnold had witnessed nothing more than natural phenomena that only appeared like flying "saucers skipping over the water" at 1,700 mph.

The decade of the 1950s saw the shift of science fiction from paperbacks and magazines to the silver screen. Flying saucers and their alien pilots were now being imprinted onto the psyches of industrialized nations in panoramic technicolor. As hundreds of thousands, and even millions of theater goers were deluged with the extraterrestrial phenomenon, the numbers of reported sightings escalated proportionately. UFO protagonists claim that the proliferation of sightings was not the result of public "overeagerness" to see UFOs, but of more eyes watching the sky.

By the 1960s, counterfeit UFOs were being mass-produced by any teenager with a dry-cleaning bag, balsa wood, and a candle.[6] High-altitude weather balloons and meteorological phenomena produced reflected atmospheric flashes that caused even the best trained pilots to believe that they were witnessing a true-to-life UFO. Many unmistakable UFO sightings were retrospectively learned to be mere test flights of conventional aircraft outfitted with special lights or other test gear. UFO debunkers such as Philip Klass began to investigate UFO sightings with the same vigor as UFO protagonists, apparently explaining most with normal terrestrial phenomena. These facts were also learned by the public, and the UFO craze gained some perspective in conventional circles.

---

5 Arnold reportedly did some writing on the subject later that tended to diminish his "disinterested" status. Based on his own experience, and those of others, he came to believe that there was something "alive" about the UFOs.

6 It is interesting how many of these inexpensive lighted hot-air balloons have appeared on the front pages of newspapers with headlines announcing the arrival of UFOs to the local community. The balloons provide impressive night sky effects, including high-speed formation maneuvering, and have fooled thousands of observers.

The Controversy Over UFO Evidence

With the affluence and technological advancements of the 1950s came the general availability of inexpensive, good quality photographic equipment. UFO reports began to be accompanied by photographs of flying saucers. Expert analysis revealed no evidence of counterfeiting in many of these photographs, though many were obvious fakes. It did not take long for people to learn that a snapshot of a Buick hubcap flying through the air, or a double exposure of eerie lights in the night sky could bring instant national fame, and sometimes fortune. While most of the genuine—at least unmanufactured—photographs reveal obscure shapes and smudges in speckled skies, a few, still frequently reprinted in UFO books, present clear images of saucer or disc-shaped objects in clear skies.

UFO Photographs

One such photograph was taken in McMinnville, Oregon, on May 11, 1950, by Paul Trent. This photograph shows a clear image of an up-side-down-plate-looking craft, with a small, pointed dome in the middle of the top. The photograph is one of the few ever to pass the scrutiny of the famed Condon Committee, an investigative committee well-known for its unkind treatment of UFO witnesses. The Condon Committee concluded, "The simplest, most direct interpretation of the photographs confirms precisely what the witnesses say they saw." Modern photographic analysis confirms that the Trent photograph is authentic. The two photographs were recently subjected to thorough computer enhancement scrutiny. The results prove that (1) the photographs contain no wires to hang a UFO model, a favorite debunker claim; (2) the UFO is an actual three-dimensional image, not flat or superimposed; (3) the UFO was at least one kilometer away; and, (4) the UFO is twenty to thirty meters in diameter. Interestingly, however, debunkers fail to accept the photograph as authentic, not because of any signs of counterfeiting, but because of the Trent family's "laid-back" attitude about the photos. This scholarly conclusion is based on the fact that a debunker had found the negatives on the floor near the sofa, where the children had knocked them off the coffee table.

Another difficulty with the testimony of the Trents is they are "repeaters." A repeater in UFOlogy is someone who has received more than his or her fair allotment of UFO sightings—generally, just one. Although the status of repeater is often the death knell of one's credibility, it is, as we see below, very much a part of the UFO phenomenon. Photographic analysis aside, there remains a question concerning the time of day the Trent photographs were taken. This, however, appears to be a post-analysis debunker fall back argument.

One of the most fascinating UFO photographic windfalls occurred in Northern Utah in 1952. Delbert Newhouse and his wife were driving near Tremonton when she spotted flashes in the clear day sky. Delbert, a military aerial observation photographer, happened to have his photographic equipment in the car with him: a 16mm hightech movie camera. When he realized that he was observing something very rare in 1952 aviation, he reached for his camera and photographed a cluster of shiny discs flying in formation through the summer sky. Government Intelligence analyzed the film for thousands of hours, and finally pronounced it to be authentic. The only other possible explanation proffered by government analysts was that the shiny objects may have been a flock of seagulls, but their own data demonstrated that the reflective nature of the objects was too radiant for seagulls, or any other type of bird, and they opted not to further analyze the theory.

The 1960s and 1970s produced a flood of UFO sightings and other close encounters, together with prolific "evidence" of UFO existence. Hundreds of photographs were published depicting clear shapes of advanced flying machines. Educators, policemen, ministers, engineers, and civic leaders reported seeing discs, saucers, cylinders, and other strange flying objects. People above reproach, former skeptics, were now confirming what they had seriously doubted only a moment before becoming witnesses themselves. Airline pilots dodged flying discs buzzing around the 550 mph passenger planes as if they were standing still. Radar centers tracked bogeys on their screens travelling at thousands of miles per hour. Reconnaissance jets were scrambled around the globe only to be chased back to base or to see metallic discs disappear into high altitudes in a flash.

Again, UFO debunkers attempt to eliminate many of such encounters as mistaken observations of bright planets, meteors, and terrestrial aircraft. It is with fervent zeal that debunkers attempt to convince the public that *all* such reports are explainable by natural phenomena. This is just not so, however, as we see below.

The Gulf Breeze UFOs

In some cases, UFOs have "haunted" certain locations for weeks, or even months, allowing the local citizens to see them many times. In one such case, citizens have been able to move in with sophisticated photographic equipment to record their presence. This is the case of the Gulf Breeze, Florida, sightings. There, in late 1987, Ed and Frances Walters, local prominent business people, as well as many others, reported numerous sightings of flying vehicles. Ed and Frances armed themselves with a Polaroid camera (it is difficult to fake Polaroid photographs), a video camera, and even a sealed four lens 3-D camera supplied by UFO investigators. In the book *The Gulf Breeze Sightings,*[7] the entire UFO experience between November 1987 and May 1988, is set out in detail, accompanied by more than twenty of the most impressive UFO photographs in existence. During the six month sighting period, 135 of the local citizens of Gulf Breeze confirmed having seen the same flying vehicles photographed by the Walterses, and published in the local newspapers.[8] Experts from Jet Propulsion Laboratories and NASA examined the photographs and a video tape for nearly a year, concluding that they were unable to find any indication of counterfeiting in the photographs, the video tape, or in the person of Ed Walters, the primary photographer. Psychologists, UFO sighting investigators, and polygraph experts have fully examined Ed and Frances[9] Walters, and concluded that there exists a low probability of a hoax on their part.[10] The only reservation, which continues to be expressed by the photo analysis experts, is that the Walterses are "repeaters."

Experts with the Mutual UFO Network (MUFON), an organization holding itself out to investigate and document reported UFO sightings, agree that the Walters case is the best documented American case to date. The author interviewed many of the investigators in the Gulf Breeze study, and all of those interviewed assert that they absolutely believe that the

---

7 Ed and Frances Walters, *The Gulf Breeze Sightings,* William Morrow and Company, New York, 1990.

8 Since the publication of the Walters' book, many more sightings have been reported in Gulf Breeze, Florida, but none so close and clear as during November 1987 through May 1988. Ed continues to photograph and video tape the UFOs, obtaining phenomenal results that have been viewed on national television and scrutinized by the best analysts.

sightings occurred, and that the Walterses were not perpetrating a hoax. In the interview of Duane Cook, the editor of the *Gulf Breeze Sentinel,* he revealed that he was quite skeptical of running the story at first, although he had the photographs and initial analyses of them in hand. The turning point for him was when he decided to ask his parents' advice, who were visiting from out of town. He expected their usual conservative high eyebrow at the thought of publishing such nonsense, when instead he was met with, "That's the thing we saw on our way into town the other night!" They had observed the same UFO as they were driving to Cook's house, but did not know what it was, and decided not to say anything about it. Of course, Cook ran the story.

The Gulf Breeze sightings produced over 60 photographs of the UFOs,[11] 41 of which were taken by the Walterses, none of which was ever claimed to show evidence of counterfeiting by photographic analysis experts. Because the Ed Walters case contains many common encounter elements, it is referred to in this chapter and those below as something of a benchmark of what appears to be occurring in contemporary UFO sightings and encounters, employed to introduce or explain some of these phenomena.

UFO Design

The UFOs sighted in the Gulf Breeze area are somewhat different than the normal metallic discs or flying saucers often reported. Interestingly, they do not exhibit that high-tech, state-of-the-art look that we often see on the covers of scandal magazines or in our own *Star Trek.* They are, instead, somewhat taller and bulkier than expected. General UFO design appears to be more functional than aesthetic—new car models have better aesthetics

---

9 With the exception of Frances undergoing a polygraph examination—she was never asked. Ed Walters reports that his examination was as grueling a two days as he has ever undergone.

10 All of this is not to say that the Walterses were not inundated with debunking efforts to assassinate their characters. Approximately two years after their initial UFO sightings, debunkers planted a model of one of the UFOs in the Walters's house, and found a young man who would testify that he was in on the entire hoax. The young man could not withstand the scrutiny, and confessed that he had been put up to the debunking effort.

11 Not all of the UFOs were exactly alike. There does seem to be similarity between them, however, as between a particular manufacturer's line of automobiles.

than do many UFOs. Besides the basic rounded and contoured shape of the main body, the Gulf Breeze UFOs have what appear to be diamond and square shaped "portals" which encircle the entire craft. The diamond-shaped portals are reported to radiate with bright light at times, while the square shapes appear completely dark. The portals are reported by some to be unequally spaced around the craft. The main body varies in reported color, a glowing "grey/blue to a rich orange-brown," sometimes changing color while being observed. Nearly all of the Gulf Breeze UFOs had a large bright light in the center on top,[12] and a "power-ring" looking light on the bottom. Ed Walters recounts that the power ring was generally "bright white with a darker orange core." This feature is prominent in the available videos and photographs.

The power ring seems to be of importance to the mobility of the UFOs, although this is merely an observation. For instance, often times the power ring would become very bright just before the UFO flashed away, or "winked out."[13] During normal operation[14] the inside of the power ring is described as a "twisting, throbbing mass of . . . silent energy storm." The photographs and video taken of the power ring support this description. In the Walters photograph #19, wherein the UFO is approximately 10 feet above the road, the bright light of the power ring is reflected off the road. Photographic analysis reveals that the light source was very bright, but experts were puzzled by the phenomenon that the light appeared to be confined to a specific area, and did not light up more of the road and surroundings. This nonradiating quality also seems to be common in UFO

---

12 This feature is nearly universal in UFO reports.

13 Observers, including Ed and Frances Walters, reported that when the UFOs left in a hurry, they did so in such a manner that left the observer wondering if the craft had dashed away at incomprehensible speeds, or if it had just "disappeared." Some of the photographs show streaking as if the UFOs had moved quickly. However, the 1 min. 38 sec. video tape shows only that the UFO was there in one frame, and gone in the next, indicating that it had instantly disappeared. Again, this characteristic is very common. A more recent Ed Walters video, however, shows a UFO accelerating instantly from a dead stop, streaking across 20 percent of the television screen in a single frame.

14 Normal operation of the UFOs was to hover or move slowly from place to place, bobbing and weaving as they went. Less frequently, the UFOs could move great or short distances in a flash.

encounters. A final interesting feature of the power ring is that it appears to be the source of a blue beam reported by the Walterses and other observers.

The Blue Beam

Ed Walters's very first reported encounter with one of the UFOs was accompanied by his being trapped in such a blue beam. After describing how the UFO, appearing like it was "right out of a Spielberg movie that had somehow escaped from the film studio," just floated into his neighborhood, over his neighbors' houses,[15] Ed Walters says he dashed for his Polaroid camera that he uses in his construction business. He recounts:

> It glided along without a whisper of sound.[16] There was no hum, no wind, not a single disturbance to the air, trees, or houses as it passed over them.[17] While rocking back and forth, it did not seem to spin, so I never saw all sides, only what was in the photographs.[18]

After Walters had taken several photographs, the UFO had moved directly over him, and as he looked up into the power ring he explains:

> Bang! Something hit me. All over my body. I tried to lift my arms to point the camera. I couldn't move them. They were blue. I was blue. Everything was blue. I was in a blue light beam. The blue beam had hit me like compression. It was pressing me firmly, just enough to stop me from moving.
>
> I screamed, with my mouth frozen half open, but the sound was hollow. Dead, like a vacuum. I couldn't even move my eyes or eyelids. I thought I was dying. I was trying to breathe, there was air, each breath shallow. . . .
>
> The best I can tell, this all took less than twenty seconds. Then my feet lifted off the ground. I screamed. A voice groaned in my head. "We will not harm you."
>
> I screamed again.
>
> The deep, computerlike voice said, "Calm down."

---

15 After the Gulf Breeze Sentinel had published Walters' photographs of this UFO, witnesses came forward with testimony that they had also seen the UFO that evening near the Walter's home.

16 In all the reports of these Gulf Breeze UFOs, the observers are unanimous that none of UFOs made an audible sound.

17 This lack of disturbance of the air is likewise significant, because many reports indicate that a UFO would flash a quarter mile in a second, just over the observer, then back to the original position in a like amount of time, with no wind, sound, or sonic boom. Again, this is a very common feature in UFO sightings. In fact, many observers report that all common ambient sounds disappear during these encounters.

18 *The Gulf Breeze Sightings,* p. 28.

But it was in my head, not my ears.

I screamed, as well as I could, "Put me down!"

A few seconds passed as I slowly rose away from the pavement. A dream? Hell no! This was real. The feeling of helplessness was the worst. No control just a piercing smell, a little scent of ammonia mixed with heavy cinnamon that scorched, then stuck to, the back of my throat.

My heart was pumping so hard I could feel its throb as it thumped against my unmoving chest. I could feel the thumping vibration pass down my legs.

The voice groaned, "S-t-o-p i-t."

I screamed, "Screw you!"

All this happened fast. Now I was about two feet above the street. I panted for air, but the smell stung my lungs. My brain started to black out, so I screamed, "Aagghh!"

The scream was black and dull, just outside my mouth. Almost the way you feel if you dive to the bottom of a swimming pool with the pressure holding everything, even your own voice, close to you.

The voice came back, but now it seemed to be female. An easy hum filled my head. Suddenly, from within my head, came the sharp vision of a dog. Then another and another. I was confused. What are these dogs? Rapid visions, one after another, on and on. It seemed that I could almost see words beneath the dog visions. Something was flashing dog pictures in my head just as if they were turning the pages of a book.[19] The hum continued. I had the sensation I was four feet above the ground.

Wham! I hit the pavement hard and fell forward on my knees. The blue light was gone. The hum was still in my head, but it quickly decreased and was gone, like the hum of a speeding car as it races by. I collapsed onto my chest into the middle of the road, filling my lungs with real air. My stomach turned and I choked, trying not to throw up.[20]

Ed Walters says that as he rolled to his back he saw a small airplane entering the area's airspace, and assumes that the plane scared the UFO away in the nick of time.[21]

If not for the photographs that Ed Walters was able to take during this sighting, it would be an encounter that he would never have reported—understandably. If not for the photographs being determined to be authentic by leading photographic experts, Ed Walters's story would be nearly impos-

---

19 In a later encounter, Ed Walters reports that the UFO attempted to distract him with physically implanted or projected flashes of naked women. Although this, in itself, could tend to detract from the credibility of an UFO sighting, in this case, it lends some credence. Ed reports that the UFO flashed indiscriminate pictures at him; naked women that are not generally thought of as "attractive." Apparently, the UFO was not aware of certain aspects of Earth culture and male preferences. Ed was not distracted.

20 *The Gulf Breeze Sightings*, pp. 29-30.

sible to accept as anything but contrived. Although thousands claim to see UFOs, Ed Walters's account is pivotal because of the photographs. There exists so little credible, tangible evidence of UFO encounters that Ed Walters's bonanza of photographic confirmation renders it a great case for analysis.

The UFOs returned many times, and made many attempts to abduct Walters. Investigators feel, after subsequent events, that Ed Walters had closer ties with these UFOs than he realized, or remembered, and there was a specific purpose for their repeated attempts to pick him up.

The blue beam described by Ed Walters was seen repeatedly by him, and by six Gulf Breeze residents in all,[22] and was photographed from a distance and up close by Walters. We learn of other properties and functions of the blue beam in similar encounters had by Ed and Frances Walters. On one occasion, Ed recounts he was scurrying out the back door to get a photograph of a UFO he spied behind his house when a blue beam flashed down at him and hit him on the leg. Rather than just having the effect of rendering Ed's leg useless or paralyzed, the beam effectively froze the leg in its three-dimensional position. Ed's momentum carried him forward, with his leg "pinned" where it was, slightly injuring him as he hyperextended the knee. Ed, with the assistance of Frances, was barely able to pull himself free of the blue beam. However, on another occasion when Frances dodged a blue beam that was near her, she reports that "[l]eaves and bits of gravel swirled around and within the beam."[23] Besides attempting to elevate Ed Walters with the blue beam, the pilots of the UFOs evidently use the blue beam as a transportation device to exit and enter the UFOs.

Telepathic Communication

An important aspect of the above-detailed Ed Walters first encounter is the apparent communication between the UFO and Walters. During many subsequent attempts to pick up Walters, he heard voices in his head coaxing him into the open, and telling him not to resist. The voice would assure

21 As observed below, UFO pilots may be surprised by unnoticed traffic entering their field of control. If this occurred here, it could be the reason the UFO abandoned its design to abduct Ed Walters.

22 Interestingly, some reported the blue beam being "shot into the water in the Gulf of Mexico.

him that he was in no danger, not from the UFO anyway, and that he should go with them. Other times, the voice insisted that he was indeed in danger, and that going with the UFO would help protect him. The voice soon began calling him by a new name, "Zehaas."[24] Ed had no idea why it called him by this name, but the voice, nearly always female after the initial communication, beckoned him, "Zehaas, Zehaas." Not only did Ed Walters hear the computer-like voice, and the later female voice, but he also heard, or apparently overheard, human voices speaking in Spanish, and sometimes in other languages that he did not recognize. The Spanish voices, evidently of an abducted Earth couple, were accompanied by the sound of a crying baby.[25] In Ed Walters's limited understanding of the Spanish language, he heard the man complain that they were provided with nothing but bananas to eat. The man observed that "they," their captors, never seemed to eat anything but bananas themselves. In addition to the voices, Ed heard ambient sounds like doors, clanks, bumps, compressed air releases, and other background noises. All of this latter communication resembled eaves-dropping, or inadvertently overhearing conversations, as if something had gone awry with the UFOs' communications systems, into which Ed Walters had evidently been plugged in some way.

In addition to the voices, Ed Walters often heard the hum in his head, like that of his first encounter, just before a sighting would occur. This is the purported reason he was able to have his cameras ready to photograph the UFOs when they appeared. The many opportunities to photograph the

---

23 *The Gulf Breeze Sightings,* p. 179.

24 Walters was contacted by a woman who explained that "Zehass" was very close to the pronunciation of the Spanish word "cejas," meaning eyebrows. He noted that he has very dark, bushy eyebrows, and sensed that the hairless UFO occupants liked them. As ridiculous as this explanation sounds, other men report a feeling that they have been selected, at least in part, because of the bushiness of their eyebrows and other body hair. They observe that they "feel" physically admired by their abductors, and that there exist certain physical characteristics of humans that these abductors intend to breed into their own species, as discussed below. "Ed," in *Intruders: The Incredible Visitations at Copley Woods* (Budd Hopkins, 1987) reports, "At that time in my life my hair was thick and coal black, and I don't know if they told me or I just had the impression that they liked my coal-black hair, and they liked my . . . they like our features. They like our skin, and they like our eyebrows and they like our hair." (*Ibid.* p. 139)

25 The presence of the baby is significant in light of the UFO reproductive experimentation discussed below.

UFOs is a point of great interest in the Gulf Breeze sightings. The UFOs allowed Ed Walters, and others, to photograph them. However, early on Ed heard the voice say in Spanish, "Los fotos son prohibido [photographs are prohibited]," just as he took another. Just then, the female voice, sounding suddenly earthly, said, "You can't expose them. They won't hurt you. Just a few tests. That's all." The implication was clear that the voice belonged to an Earth woman, acting in concert with the UFO pilots.[26] However, the plea not to expose the UFOs is contrary to the overwhelming evidence that they wanted Ed Walters and others to photograph them.[27] This inconsistency is one that has not escaped any of the investigators, or Ed Walters. Ed Walters muses, "if they want people to know about them, why don't they just land in the middle of the Orange Bowl?"

Ed Walters is not the only person in the Gulf Breeze area to claim that he heard voices from the UFOs. *The Gulf Breeze Sentinel* reported that a father, mother, and their adolescent son watched an approaching UFO for several minutes, and tried following it when the boy suddenly "overheard" a male voice give the command, "Stop!" No one else heard the voice but it was very clear to the boy. He was quite shaken and became ill following the experience. The *Sentinel* also reported that another woman, with her grown daughter, observed a UFO for an hour, most of which time they lay on a bed in their house in fear for their lives. The woman reported that during the entire time, she heard a thumping that sounded like a loud heartbeat. Her daughter heard nothing. The woman insisted that she had the terrifying feeling that the "ship" was watching her during the episode.[28] The "hearing" of voices during UFO sightings is a common theme in the

---

26 Many who have reported seeing UFO occupants report that they are sometimes accompanied by Earth humans, or hybrids—half alien, half human.

27 Although they wanted some to see them, the UFOs appeared to go to great lengths to ensure that certain others did not see them. Investigators soon discovered that the UFOs disappeared instantly when Ed Walters gained knowledge of the investigators' presence. Sensing that the UFOs were "plugged into" Ed's sensory perceptors, the investigators withheld information from him, enabling them to witness the UFOs themselves on a couple of occasions.

28 Duane Cook, then the editor at the *Gulf Breeze Sentinel,* confirms that all of the reports attributed to that newspaper appearing in Ed Walter's book are genuine reports.

UFO phenomenon. Persons contacted by these voices are often given instructions concerning imminent encounters, e.g., where to go and what to do.[29]

Many of the events of the Gulf Breeze sightings are consistent with other sightings all over the world. Multiple sightings of different kinds of craft, communications with the observers, and similar phenomena are reported weekly by persons not known to fabricate stories. Governments appear unilaterally, and uniformly, to deny that any such phenomena exist, while, as in the case of the Gulf Breeze sightings, they go to great lengths to monitor the activity and cover it up.

Government Coverup—Men In Black

The stories of government coverup attempts abound with almost every UFO report that surfaces. By some reports, the government is as in the dark as anyone concerning UFOs, and by others, the governments of the world are active participants in a massive coverup. In many UFO sightings in which evidence of the sighting is claimed to exist, government "men in black" are reported to appear at the witnesses' doors demanding the evidence and threatening witnesses in the event they divulge any information regarding their sighting. If such evidence is turned over to these men in black ("MIB"), it is never seen again. Many UFO witnesses also claim that their houses and cars are buzzed by enigmatic helicopters following their sightings.

Even more interesting are the "spooky" aspects of MIBs. For example, witnesses claim to see a UFO on their way home from shopping and when they arrive at their house (within five minutes), MIBs are already waiting for them, making demands and threats.

During the author's investigation of MIB activities in the Salt Lake City, Utah area a now middle-aged woman of conservative Christian background reported the following incident. About twenty years ago she caught a glimpse of a UFO and mentioned it only to her parents. Within a day, she and her parents received a visit from two men representing themselves to be from the U.S. government. The men were grave and threatening, and

29 Kathie Davis (really, Debra Tomey) and Betty Luca, abductees discussed below, also heard a voice within them call them by name just before their abductions.

told the small family that they had better never disclose the details of the sighting. The three were perplexed because the sighting had been so brief, and they had told no one about it, so they wondered how these men knew about it.

A couple of years later, the young woman needed employment and was thrilled when she received an offer of a government job over the telephone. She was told that she would be picked up for testing the next morning. A plain looking van pulled up in front of her house the following morning, quite early, and she and others were driven through the desert west of Salt Lake City to an abandoned-looking building. There, she and the others were "tested" for the entire day and into the night. The testing, however, had nothing to do with typing or filing skills. It was centered in psychic and related abilities of those being tested. After the long, grueling day of such testing, the young woman was returned to her home, and eventually gave up hope of being given the "job" she had been offered.

The MIBs are often seen driving showroom quality cars that are twenty- or thirty-year-old models. Further reports claim that witnesses have attempted to follow MIBs in their unique automobiles, only to come to the end of a cul-de-sac or other dead end road, and find no sign of the MIBs or their car. Some investigators feel that the MIB and unusual helicopters are so enigmatic that they may actually be connected with the UFOs instead of the government.

Raymond Fowler, UFO witness, researcher, and author, relates a cross-country ski outing with his daughter during which they both saw a man dressed in black standing under a tree in the distance. Fowler recounts that he had a slightly uneasy feeling about the man, possibly because the man was in an isolated area surrounded for miles by deep snow, dressed in city clothing. They watched the man periodically as they neared the tree. As they approached, their attention was momentarily diverted. When they looked back to the tree, the man was no longer there. They skied over to the tree, only to find it surrounded by deep snow and no footprints.

UFOs Around Energy Installations

Although less documented than some citizen sightings, certain energy installations have at times been the object of government concern with military reports of UFOs in the vicinity. The Kuwaiti oil field installations were at one time the object of multiple UFO sightings, causing the Kuwaiti government some distress about its most precious resource. Often, high voltage power stations or transmission lines have been the source of UFO sightings. UFOlogists proffer guesses at why UFO pilots might be interested in such low-tech power supplies, but nothing substantive has yet come to the fore. Some UFO debunkers feel that high voltage power supplies may actually be the sources of some UFO sightings. They theorize that certain "fireball" sightings may actually be energy flashes that emanate from high voltage sources in a manner not yet identified by electrical science.

Gentle Invaders

The 1977 motion picture *Close Encounters of the Third Kind* attempts to synthesize many common UFO sighting patterns and themes into a few representative sightings and encounters. Small space vehicles emerging from larger "mother ships," bizarre effects on spacecraft observers and electrical equipment, concentrated light beams that probe, disrupt, and sunburn are all common themes in UFO annals. What is uncommon about the motion picture is the benign nature of the aliens. Although the television series *Star Trek* had portrayed certain aliens as benevolent, while others as savage and imperialistic, gentle invaders were veritable newcomers to the silver screen, with a few exceptions. *Close Encounters of the Third Kind* is instructive, however, for those who seek a "*Reader's Digest*" condensed view of what hundreds of UFO books reveal about UFO sightings and related encounters.

From Hardware To Soft Sightings

Many UFOlogists are turning from "nuts and bolts" theories of UFO origins to decidedly more metaphysical explanations. Aside from the many paranormal explanations, some theorize that UFOs are physical manifestations of electromagnetic phenomena created by stress fracturing of the Earth's crust, much as the lights that reportedly accompany or precurse an earthquake. Some studies indicate that there may exist a correlation between seismic activity and UFO sightings. Other studies even suggest a relationship between the phenomenon and the observer, where the observed light responds to the thoughts and emotions of the observer.

These electromagnetic fields, whatever their source, theoretically generate perceptual and reasoning difficulties in persons exposed to them, triggering irrational perceptions of the lights. Supposedly, one's shaded comprehension of the nature of the light being observed is generated by one's own cultural expectation of what enigma or mystical manifestation is occurring. For instance, protagonists claim that certain British haunts well known for their UFO sightings were formerly well known for their spook lights, spirits, demons, and Madonna sightings. They speculate that two persons with quite diverse backgrounds viewing the same phenomenon at the same time will perceive something altogether different.

Spherical Probes

The subject of ball lightning or travelling lights takes on new perspective with certain of the recent UFO reports recounting encounters with "probes" that have the appearance of floating spheres of light, just larger than basketballs. Many of the sightings reported on television programs are accompanied by spherical lights that probe the area, and then return to the UFO before it leaves the vicinity. One such report is documented by Budd Hopkins in his book *Intruders: The Incredible Visitations at Copley Woods.*[30] *Intruders* tells of the encounters of a young mother, Debra Tomey, called Kathie Davis for the purposes of the book, who was plagued by repeated visitations by UFOs. Although her encounters fit more properly into the abduction discussion below, the spherical light probe is of interest at this point. Earlier in the evening, Kathie's mother reported seeing the light through the kitchen window, near the hummingbird feeder in the back yard. Significantly, within a couple of days, the shrubbery in the immediate vicinity had withered. Upon investigating the floating white lights, Kathie, during hypnotic regression, reported that she became semiconscious and her will was taken over by an external force. She was led to the backyard where a white sphere of light "looked at her" at eye level, then moved slowly downward. Suddenly, the sphere entered her, causing her

30 A motion picture made for television based on *Intruders* was released in mid 1992. The television movie was a dramatization of the kinds of events in the book, in which component characters of abuctees and investigators were created to experience the UFO/abduction phenomenon.

much pain and flooding her body with light. She said that she felt as if a bolt of lightning had hit her in the chest, then flowed throughout her body.[31] She was unable to see most of the time because she was flooded with the light, as if lit up internally. She knew that these "globes" were connected with a UFO because she saw them detach themselves from the outside of a UFO and float around "exploring."[32]

Modern Airships

"Paul" says he did not believe in UFOs despite his own brother's testimony of having seen one while traveling on a commercial airliner. In June of 1964, Paul became a believer.

Paul and his wife had moved up Emigration Canyon, just east of Salt Lake City, Utah, and were visiting at their neighbor's home. Paul's wife had to go to work the next morning so she excused herself and went home around 11:30 p.m. He stayed until around 1:30 a.m. His hosts were completely exhausted from their fun, so he showed himself out the side door of the house. There was no light on that side and the sky was perfectly clear. He began to "feel" a noise or vibration, that pulsed through him like an electronic signal or hum. In that part of the canyon one could not make any sound at night without setting off the entire dog population of the neighborhood, but as he listened and the pulsating tone became louder, there was no barking. A power transformer had burned out not long before, so his first thought was that it was going out again. He looked, but saw no sign of fire up or down the lines.

"Then this object came over the ridge behind our home, blocking out the stars" where it passed, as if an opaque silhouette was between him and the starry sky. He could see no details, only the opaque outline where the stars had disappeared. He estimates the length of the object as being the width of a football field, approximately 150 feet. He says, "It looked like a dirigible," but insists that it was not. It seemed to be travelling about 30 or 40 mph as it crested the ridge, then slowed down. It "floated" toward

---

31 *Intruders,* at page 43.

32 Betty Luca reports several encounters with such probes also as documented in *The Watchers: The Secret Design Behind UFO Abductions,* (Raymond E. Fowler) Bantam Books, New York, 1990.

him, within a very short distance and completely stopped, just above their driveway. It was very close—30 feet away, possibly. It had round windows like portholes. He describes it as something between a "skinny football or a fat cigar." There was a pale green light inside.

"I had a frightened feeling I felt like a rabbit hiding behind a bush." He wondered how they could see him in the perfect darkness, "yet they stopped and they came to the portholes, and they stood there, and I could make out shoulders and the forms of heads, but I couldn't make out any hair or anything like that." As the "people" came to the portholes to look, he noticed lights underneath that had the appearance of "welding torches." The lights were bright and filled with different colors.

At that time Paul received what he felt was a "telepathic suggestion" from the people in the UFO. He could not make out discernable words, but "felt" an invitation to come with them. The thought of his "pretty young wife and children" coupled with his immediate fear caused him to reject their offer, attempting to convey his feelings of responsibility toward his family. They just "continued on," much as if a helium balloon had been released.

He believes the entire encounter lasted just a few minutes. He ran into his house, and shook his wife to wake her up, to which she responded "you're dreaming, go back to sleep." All of the animals remained silent and out of sight following the encounter he felt that they were aware of what was happening, and had been too frightened to come out after that.

Paul worked for the Utah State Department of Highways at the time, so he telephoned the Civil Air Patrol and asked if there were any dirigibles in the area. They informed him that there was nothing like that within several states, and assumed that there were no lighter-than-air vehicles being used anywhere in the United States around that time. He proposed many possible scenarios and explanations to the Civil Air Patrol that may have explained what he had seen, but they knocked them down one at a time, finally becoming irate, and demanding his name. With that, he hung up the telephone.

Paul says, "I felt extremely inferior to whatever it was that was up there observing me." He adds that there are various levels of apprehension that he feels in different situations. In this case he says, "I would say that I had a greater level of apprehension concerning this than I would about finding a rattlesnake under my seat." However, he is adamant about feeling that the UFO occupants were completely benevolent and meant no harm.

Mother Ships

Some witnesses report the sizes of UFOs to be "as big as a football field." Some claim that their UFO was larger than a football field, or that it filled the sky, as in the case of Steve and Dawn Hess, whose encounter is related below in chapter 7.

In the mid-1960s, "Carl," who was eleven years old, was in the alfalfa fields of his family's farm with his seventeen-year-old sister, doing the watering. Carl's family lived in Howell, Utah, about fifteen miles northwest of Tremonton. Their 350-acre farm was set in a valley, about five miles wide and eight miles long, running north to south. The family farm lay in the middle of the valley, and in the south end of the valley was Thiokol plant number 78, part of a very large Aerospace firm.

Carl and his sister were doing the watering at about 2:00 a.m., admiring the clear sky with its bright moon and stars. They were struck by the sudden appearance of a craft in the north end of the valley as it came over the mountains, about three or four miles away from them. It was enormous, filling almost one-third of the valley from side to side. Carl and his sister believe the UFO was about a mile in diameter. The craft was flying low—between 70 to 100 feet from the ground, they recall. It moved slowly and steadily from the north end of the valley. It made no noise at all. The two youths did their best to duck down in the alfalfa, but it was only a foot tall at that time. They jumped into an irrigation ditch that had not yet been flooded. Carl's sister reports that she prayed the entire fifteen to twenty minutes it took the craft to travel the length of the valley—prayed that it wouldn't notice them, or if it did, that it would not molest them in any way.

Carl's only memory about what he was thinking at the time was how big the craft was, how silent, how overwhelming—and, how scared he was. Carl and his sister describe the craft similarly—a double convex ("like two curved plates stuck together") with a ten-foot-high mid-section separating the plates. The belt-like section around the middle had evenly spaced lights around the entire perimeter, like large holes in a belt. Carl's sister recalls that these round lights were as large as the mid-section, ten feet in diameter, and yellowish-orange in color. The lights were so bright that as the UFO came over the mountains it lit up the entire valley. Although the craft took nearly twenty minutes to pass the eight miles through the valley, and was only a mile or so from them as it passed, Carl and his sister had difficulty making out any distinguishing characteristics other than the basic shape. It was more of a silhouette surrounded by the bright yellowish-orange lights, although it appeared to be completely solid and tangible in every respect. They could see that the craft was "smooth, curved with a dome," but could see no other appendages or irregular shapes attached.

The craft "floated" from north to south, making no noise at all, and continued without any deviation from its course or speed. Carl could not believe the silence associated with such a gargantuan structure. There was no wobble, no wind—nothing. It just floated in a straight course, very slowly. Carl's sister says it was "eerie."

The only change occurred when the UFO reached the south end of the valley. When it arrived at the Thiokol plant, it stopped, and hovered over it. After that, the yellowish-orange lights changed their hue to a bright red, and the craft drifted up over the mountains at the south end of the valley, then flashed upward and out of sight in just two or three seconds.

This was Carl's only close sighting. It was not his sister's only sighting, however. When she was younger, she and her brother (between her and Carl in age), and some neighbor children were walking outside late one evening. They were near each other, and suddenly saw three strange lights in the cloudy night sky. One light separated from the others and approached the children. As they watched, they could see that it was a saucer-shaped UFO. When the craft was only twenty feet from them, directly overhead, the children saw a panel slide open. From the open panel they saw a tubular light beam telescope its way to the ground. They instantly found themselves surrounded by bright light as they huddled together, seeing how they were lit up. The light flooded them for three to five minutes as they felt that they were being examined or that their thoughts were being read. Carl's sister was especially concerned that because she was the oldest of the children, it was her responsibility to protect them. She assumed that although her father was working in an outbuilding on the farm, he was over a mile away, and would not hear their cries if they called for help.

As the children stood there huddled in the light beam, it never occurred to them to run. Carl's sister reports that her thought processes seemed to function normally, although she was terribly frightened, but she never thought to run the entire time. She says that there was no apparent physical abnormality about the light beam. It was not hot or cold, and she did not feel paralyzed or controlled. They just assumed they were being watched by someone with a much higher capacity than we have.

After three to five minutes in the light beam, the children saw it begin to retract. It telescoped up a ways—again, as though it were a solid conduit of something—then it "gathered itself and retracted into the UFO." When the light beam had reentered the UFO, the panel slid closed and the UFO joined the other two, and together they flew off into the clouds. This was another no-noise encounter.

This was not Carl's sister's first sighting. Years before she had spied a "shiny basketball-looking" sphere above a neighbor's house. She watched it for twenty minutes as it floated around the house. Although she kept calling for her mother or someone else to come and look at the shiny sphere, her mother could not get away from a chore and no one else was old enough to care. It took off in a great hurry, disappearing in the distance in a second or two.

Carl's older brothers had observed UFOs with their father on several occasions while working the farm at night. The family spoke often and openly about the UFOs that were visiting their area. Carl's father also worked at Thiokol for 28 years. It is widely reported throughout the world that UFOs are often seen in the vicinity of aerospace facilities. I asked Carl if his father had ever reported these UFO sightings to his superiors, or if Thiokol had a policy requiring the reporting of such anomalous encounters or observations. Carl did not know, and his father died a few years ago. Carl's sister, however, says that she knew several Thiokol employees who had seen UFOs in the area, and knew that they had reported them. Carl's sister relates the story of her friend who had an unusually close encounter in the valley. She was driving along in a pickup and watched a UFO come right down over her. She was especially surprised when the UFO landed right on the roof of the cab. With that, the pickup's engine died and all of the electronics went out. She experienced a blackout of her own, and when she regained consciousness the UFO was gone.

We see in Carl's account many of the classic UFO themes: an area frequented by UFOs; a family with multiple encounters; the classic double-convex craft with yellowish-red lights around the circumference of the hull;

the UFOs moved without any sound, floating along in quiet observation; and rapid acceleration at unfathomable speed. What was different about the sighting, of course, was the mammoth size of the craft. Many observers report UFOs the size of football fields, or three times that size. These are thought to be very large by UFOlogists. A UFO approximately one mile in diameter, however, is a very rare sighting indeed. One staggers at the principles of science that allow such a craft to hover silently in Earth's atmosphere, disturbing nothing, creating no wind or other interference. Even more incredible is the propulsion system that propels such a monstrosity from 200 feet above the ground to high orbit in a brief second or two without a whisper of a sound, without a rush of displaced air, without a sonic boom, and without over-heating the craft's hull.

Yet, the only difference between Carl's sighting and those of tens of thousands of others is the size of the UFO. Whatever principle is at work here fails to differentiate between ten-foot wide craft and those a mile wide.

4
Saucers On The Ground

Landing Sites

With the frequency of UFO sightings that were pouring into local police stations and newspaper offices in the last half of this century, we would be surprised if no account ever reported that a UFO had landed on the Earth. Science fiction literature and motion pictures provided an incessant parade of such landings; the public assumed that was the purpose of alien visitation, if such existed. It was not long before witnesses around the globe began reporting that UFOs had been seen on the ground.

By the 1960s, UFO protagonists had organized themselves into research and investigation teams, poring over photographs and interviewing witnesses in search of credible evidence that UFOs exist. With incoming reports of UFO landings, researchers and teams were scrambled to landing sites to investigate and gather data. Government investigators had also been assigned to look into the UFO phenomenon—if new technology did exist, terrestrial or otherwise, it was a basic function of government intelligence to seek it out. With these massive investigations being conducted by forensic scientists and other skilled observers, we would expect a great deal of evidence to be produced at UFO landing sites. Had murders occurred at the sites, incredible amounts of microscopic and other evidence of the presence of certain persons and equipment would easily have been produced and analyzed. However, the data have been scanty and inconclusive.

Again, as with photographs and other evidence of UFOs, landing sites sometimes have been counterfeited by publicity seekers and overzealous "believers." These practices have strained public and professional faith in UFOlogy, casting dark shadows over the authenticity of all alleged landing sites. However, in addition to the eyewitness accounts of actual UFO landings, some residual evidence exists. Most alleged landing sites are made up of some sort of circular pattern etched or charred in the vegetation, or of evidence of landing gear, generally three in number, having left their mark. The circular patterns are often composed of flattened grass or other vegetation that is often dead or dying.[1] There are occasionally reported residual signs of a chemical or radioactive substance, and some circular patterns glow in the dark for a few days before fading. An anomalous landing site existed during the Gulf Breeze sightings. A report of an analysis of the site made by Max E. Griggs of the University of Florida is reprinted on pages 343-44 of the Walters book. The analysis is comprised of testing for several possible causes for the presence of the dead grass within the circular pattern, which happened to be the approximate size of the reported spacecraft. The experts tested for chemical agents, signs of lightning strike, fungal, bacterial, viral, pathogen, disease, insect infestation, and nutrient deficiency, all to no avail. The report concluded that although no cause was found, "evidence supports either the influence of short-lived toxic chemical or exposure of the grass to an energy source capable of killing it. Either of these would have required mechanical precision to do what was observed on the field."

A similar landing site is described in the Kathie Davis case in *Intruders.* Budd Hopkins provides a series of photographs of the site ranging from six weeks following the UFO landing to one year after. A copy of the "Physical Laboratory Report" is reproduced therein, revealing that grass and soil within a couple of inches of the landing site UFO pattern are normal, while samples taken from within the pattern are discolored and quite dense. To duplicate the coloring and texture of the site samples, adjacent samples had to be heated to 800 degrees Fahrenheit for six hours. This procedure alone could not duplicate the density of the site samples, however. It is no surprise that the photographs reveal a distinct decline in vegetation within the landing site pattern, given the changed condition of the soil.

Crash And Retrieval Coverups

One frequent excuse by UFO enthusiasts for the absence of landing site evidence is removal of UFO traces by government coverup teams. Not

---

1 However, there do exist the much debated circular patterns found in crop fields throughout the world. British authorities have conducted tireless research to discover the source of these precise patterns. Some indications are that they are not man made—others claim that they, indeed, have created the patterns as a hoax.

only do UFO protagonists claim that government teams move in on landing sites quickly and remove all residual signs of alien visitations, but persistent rumors circulate throughout UFOlogy that governments of the world have recovered at least one crashed UFO and have sequestered it. A number of books, and even a motion picture or two, have represented that military bases in the southwestern United States, and elsewhere around the globe, are collection sites for such finds.

"Crash and retrieval" stories date back to the 1884 American West. The tales are similar no matter where or when they supposedly occurred. The common scenario is: there was a crash of a UFO; someone arrived on the scene to find wreckage, and often, dead or dying alien bodies; the government threatened the local citizenry not to reveal what had happened; and government personnel transported the booty to a local airbase hanger. Once safely sequestered, no civilian, military or governmental person ever reveals any portion of the story, except an occasional unnamed mole, whose story can never be checked. These crash and retrieval stories persist, even though little evidence of them finds its way to the public. These reports are difficult to accept at first glance. The notion that our government, or most other governments for that matter, is capable of keeping such events from the public strains our collective perception of governmental capabilities. Our government is totally incapable of keeping its own political leaders from revealing our nation's uppermost military secrets to the world press on the capitol steps. A successful five-decade government coverup of captured UFOs seems unlikely under these circumstances. It is further evident from our current state of technology that no technological quantum leaps have occurred as a result of discovered alien propulsion systems, or other advanced devices. We still burn fossil fuels in internal combustion engines that are reminiscent of century-old prototypes, and we clumsily blast payloads into Earth's orbit and beyond with *rockets,* not unlike those used in China centuries ago. The story is the same in all scientific disciplines—metallurgy, chemistry, and even astronomy—slow, steady-as-she-goes progress despite desperate need of quick applications of superior technology.

The absence of evident technological exploitation of downed spacecraft does not necessarily debunk the possibility of the existence of government held alien spacecraft. It is possible that such advanced technology is either too advanced to be copied with today's state-of-the-art equipment, or too damaged to be of any beneficial use. A third possibility is echoed by many who claim that the U.S. government actually possesses superior technology that it hides from the public—a secret government with secret technology.

The Roswell Crash

The best-documented crash and retrieval case is the 1947 Roswell, New Mexico incident.[2] Just before 10:00 p.m. on July 2, locals sighted what they reported as a large glowing object flying at high speeds. Later, during a thunder storm, rancher Mac Brazel heard a thunderous explosion that was much louder than the thunder claps of the storm. The next morning he found some unusual debris that he described as very thin, pliable, and extremely strong. Other reports relate that he also found a downed saucer.

Brazel called in Army Intelligence from the Roswell Army Air Field. Five days later Lieutenant Warren Haught told the Roswell Daily Record, "The many rumors regarding the flying disk became a reality yesterday when the intelligence office of the 509th Bomb Group of the Eighth Air Force, Roswell Army Air Field, was fortunate enough to gain possession of a disk through the cooperation of one of the local ranchers and the sheriff's office of Chaves County." Because the airbase was inundated with inquiries about the disk for the next several days, Brigadier General Roger Ramey made a public radio announcement that the whole affair had been a misunderstanding, and that the wreckage was merely a downed weather balloon. Two news conferences were held. At the first, photographers were allowed to view and photograph unusual wreckage materials from a distance. After the government's retraction, the press complained that they had not been allowed to see and photograph the wreckage from up close, and demanded that the airbase allow them to view the wreckage again. This time, the press complained that the wreckage had been switched, and was not what had been shown on the first occasion.

Author Whitley Strieber documents subsequent attempts to publicize a government coverup of the Roswell incident:

> Major Marcel had in 1979 contributed a videotaped statement to a documentary entitled *Flying Saucers are Real.* In his statement, the major was absolutely unequivocal.

---

2 *See, e.g.*, Charles Berlitz and William L. Moore, *The Roswell Incident* (New York: Grosset and Dunlap, 1980).

> "One thing I was certain of," he said, "being familiar with all air activities, was that it was not a weather balloon, not an aircraft, or a missile . . . . A lot of the little members had little symbols which we were calling hieroglyphics because they couldn't be read . . . . It [the metal] was not any thicker than the tinfoil in a pack of cigarettes, yet when I tried to bend it, it would not bend . . . ." Major Marcel added, "The reason that this story has remained hidden from the public for over thirty years is that General Rainey released a cover story at that point." (The General claimed that the debris had been identified as coming from a crashed weather balloon.) The Major's statement was made after he had retired from the air force with an honorable discharge a few years before he died.[3]

"Unsolved Mystery" types of television programs have attempted to document these cases, as well as some others. Again, some evidence of a crash and retrieval exists, but nothing conclusive has yet been produced. Few doubt that something was recovered in the Southwest—but what, we do not know.

Area 51

Much attention has recently been given to reports of UFO-type U.S. Military craft that have been seen floating and darting in the Nevada desert near Las Vegas. Although these reports are as unconfirmed as other UFO sightings, UFOlogists are buzzing with reports that the U.S. Government, at least the Air Force, not only retrieved crashed UFOs and their occupants, but entered into a "treaty" with aliens providing that in exchange for returning the materials and bodies and for looking the other way in abduction cases, the extraterrestrials would supply advanced technologies—technologies that are now being utilized by the Air Force, and kept under wraps.

Although the sources of these kinds of tales are unconfirmed, one former government employee has come to the fore with a very detailed account of his employment at what is known as S4. Area S4 is a highly secured area about 15 miles south of Area 51, a government research base where high tech weaponry is developed. All of this is located about 125 miles from Las Vegas, directly north. Robert Lazar claims that he was one of many physicists employed to "back engineer" alien craft in the possession of the

---

3 Whitley Strieber, *Transformation: The Breakthrough,* Beechtree Books/William Morrow, New York (1988).

U.S. Government. Back engineering is the process whereby an engineer will dissect an already existing product and attempt to copy it. Lazar says that Area S4 contains nine secluded hangers in the side of a mountain, protected by ground-to-air anti-aircraft missiles. The hangars contain at least nine alien disks.

Lazar describes one disk on which he worked, which he saw fully functional in flight, and in which he spent time while it was energized. He describes this particular model as being about 16 feet in height and about 40 feet in diameter, containing three interior levels. The bottom level contains the propulsion system, which consists of "three gravity amplifiers and amplifier guides," which are used to "amplify and focus the Gravity A wave" generated in the "reactor." The small reactor is located in the center of the middle level, and acts as a particle accelerator which bombards Element 115 with protons, turning it into Element 116. The resulting Element 116 has a half-life of a fraction of a second. When the Element 116 immediately breaks down, it releases antimatter, the heat from which is generated into electricity. The decaying Element 116 also releases the Gravity A wave, which is amplified and utilized by the gravity amplifiers and guides.

The area of the reactor also serves as the control center, having seats that are adapted to a person the size of a child. Lazar claims: "The walls of the center level are all divided into archways. At one point in time when the disk was energized, one of the archways became transparent and you could see the area outside of it just as if the archway was a window. After the panel had been transparent for a while, a form of writing, which was unlike any alphabetic, scientific, or mathematical symbols I have ever seen, began to appear on the transparent archway." Lazar continues: "I was never given access to the upper level of the disk, so I can't enlighten you as to what the porthole areas are—other than I can assure you that they are not portholes."

The gravity amplifiers create (access and amplify) an intense gravity field around the disks, which in turn creates a space-time distortion around them. These space-time distortions enable the controller to coordinate warps in the fabric of space-time, distorting it to a point that distant coordinates in the linear space-time continuum are brought together like folds in a cloth. When the target location is warped to within close proximity of the disk, the gravity field is decreased, space-time is released and allowed to return to its natural form, and the leap is made. Thus, incomprehensible distances are travelled, not in a straight line, but in a "shortcut" leap, like a secret passageway shortcut between two points.

Lazar explains that when the disk is in close proximity to a planet, which radiates its own longer Gravity B waves, the gravitational field around the disk can be amplified to the point that the space-time distortion is so intense that the disk appears to fade from view. In this phase, all an observer can see is the sky surrounding the craft. Lazar claims that three projects were being conducted while he was at Area S4. His was the gravity propulsion project. The second project was called "Project Sidekick," which "dealt with a beam weapon that had a neutron source and was focussed by a gravity lens." The third, called "Project Looking Glass," dealt with the physics of seeing back through time. Lazar says that he learned of the other projects, and the general background of the isolated effort as part of his indoctrination. Therein, he would be taken into a room with a small table and chair and given a stack of briefings to read. The briefings contained information "relating to aliens and alien technology." The aliens purportedly came here from the Zeta Reticuli 1 and 2 star systems. They come from the fourth planet out from star number 2 (binary star system). The beings are described as the small grey variety—like those encountered by Ed Walters.

Lazar could not tell from the year-dating system when the beings first arrived, but they had been visiting Earth for a long time. They had presented photographic evidence of their visits that they contended was over ten thousand years old. Technology and information exchanges continued in Nevada until 1979 when some kind of rift developed. The beings left, leaving their high tech equipment behind. This is when the back engineering program began. They will return at the date 1623, whenever that is. Lazar was hired in 1988 to replace one of several engineers killed while opening an antimatter reactor.

According to the briefings Lazar read, the alien beings shared information with our government concerning the ability to remotely anesthetize the human brain and body. This can only be done when the brain is in a relaxed mode—in or around sleep. Any stimulants like loud noise or stimulant drugs render the process ineffective. The beings credit themselves with genetically manipulating our species from lower animals, and refer to us as "containers," although Lazar has no idea "what we're containers of."

More Government Coverups

As discussed above, many UFOlogists claim that the government is covering up the existence of UFOs—not only the American government, but all governments. This is, again, an unprecedented bit of international cooperation which tends to tarnish the theory all by itself.

Why do these world governments cover up the existence of UFOs? Some believe it is done to spare their citizens the psychological trauma of learning that we are not alone in the universe. This can hardly be a valid excuse among the industrialized nations, seeing that most of their citizens already believe in the possible existence of extraterrestrial life in some form or another.

Other UFO protagonists proffer the "warm up" theory—that governments are collectively allowing the people of the Earth to slowly get used to the idea that alien life may exist, and then they will one day tell us all about it—when they believe we are ready. This theory is flawed for the same reasons as the former. Other UFOlogists propose that the aliens themselves are conducting a warm up campaign; slowly making their presence known, building up to a worldwide announcement. However absurd these theories may sound, there is, nonetheless, a place for them in New Age thinking, which is discussed later in the book.

A Documentary Smoking Gun

UFO protagonists point often to documents unearthed pursuant to the Freedom of Information Act that reveal a keen government interest in UFOs. These documents, however, contain no real "smoking gun" admission of knowledge or conspiracy. If, as is claimed by protagonists, the American government regards UFOs to be such a high level security classification, it is not surprising that no conclusive evidence exists in the released documents. The government has the option of retaining those documents it feels are still important to the national security.

Raymond Fowler evidently managed to procure a copy of a Top Secret memo discovered in the effects of the former Superintendent of Radio

Regulations Engineering for Canada's Department of Transport, Wilbert B. Smith. Smith headed "Project Magnet," a subdivision of Canada's "Project Second Storey," its UFO research program. Although the files remained classified following Smith's death, his widow unwittingly allowed his home files to be released to researchers. In his files they discovered a Top Secret document that contained the following:

> I made discreet enquiries through the Canadian Embassy staff in Washington who were able to obtain for me the following information.
>
> a. The matter is the most highly classified subject in the United States government, rating higher than the H-bomb.
> b. Flying saucers exist.
> c. Their modus operandi is unknown but concentrated effort is being made by a small group headed by Doctor Vannevar Bush.
> d. The entire matter is considered by the United States authorities to be of tremendous significance.

I was further informed that the United States authorities are investigating along quite a number of lines which might possibly be related to the saucers such as mental phenomena.[4]

Government Disinformation?

Whitley Strieber relates that a noted Documentary filmmaker approached him, having heard about his personal experiences with "visitors" much like those discussed in subsequent chapters, and wanted to compare notes. The filmmaker, who requested anonymity, reported the following.

> In the early eighties the filmmaker had a very unusual encounter with a man who identified himself as a member of the air force. Their meeting was held at an air-force base in connection with the documentary the filmmaker was preparing. He was allowed to read a briefing paper concerning crashed disks and retrieval of bodies of nonhuman beings. The filmmaker was specifically informed that he was being shown this paper at the direction of superior officers . . . .
>
> The typed pages which he had read were titled "Briefing Paper for the President of the United States." There was no specific president mentioned, and he didn't remember a specific date. He was not allowed to take notes on the spot, but he recorded his recollections later in detail. The controversial document surfaced in 1987, and the filmmaker's memory proved to be accurate.
>
> The paper described a series of crashed UFO disks at Aztec and Roswell, New Mexico, at Kingman, Arizona, and a crash in Mexico. Nonhuman bodies had allegedly

---

4 *The Watchers,* p. 186.

been taken from the craft and had been examined in laboratories. The creatures were described as about four feet tall, grayskinned, and hairless, and having large heads compared to their smaller, thin bodies. Their faces were flat without ears or nose, and had a slit for a mouth. They had large eyes. Because of their skin color they were referred to as "grays."

The paper also described direct contact between government officials and a survivor of one of these crashes. This being was called Ebe, an acronym for "extraterrestrial biological entity." The officials were told that the gray beings had carried out a long-term intervention in human affairs, manipulating mankind's biological, sociocultural, and religious evolution. The being had eventually died of unknown causes. The paper outlined the government's efforts since the 1940s to ascertain the origin, nature and motives of the beings, and, presumably, to gain some sort of control over the situation, or at least some insight into it.

The agent told the filmmaker that he was being shown the document and given the information because the government intended to release to him several thousand feet of film taken between 1947 and 1964 showing crashed disks and extraterrestrial bodies as historic footage to be placed in his documentary.

He never received the footage. Had he been shown a real document, or was he the victim of some sort of complicated disinformation scheme? When the promised footage didn't materialize, the company he was working for became disillusioned and dropped its plans for the documentary, which he believes was the real outcome desired by the air force.[5]

Whitley Strieber implies that the entire scenario was a government setup designed to raise the production company's expectations about the proposed documentary, in the hope the company would lose its interest when the promised blockbuster materials were not forthcoming. When the briefing document surfaced it was analyzed by experts who pronounced it authentic—as to its origin, but not its content.

So why would the government show such a paper to the filmmaker? The reason proffered by Strieber appears valid, assuming that the paper's contents were manufactured—which assumption is a safe one. The second question is more interesting: why is the Air Force so interested in debunking the UFO phenomenon? Observers assume that it has been a matter of pride due to the Air Force's lack of ability to control our airspace—its primary justification for existence. Others believe that the "alien alliance" treaty motivates the government to cover up alien existence in this manner. In light of these facts, one wonders if the Canadian government was not also a pawn in a disinformation campaign in the

---

5 *Transformation,* p. 116.

above-cited Wilbert B. Smith document. We could ask the same question about Robert Lazar. Reports abound of such disinformation campaigns.

If There Is A Coverup—Why SETI?

Not everyone believes that the U.S. government is involved in coverups or secret alien alliances. The fact is that the government has conducted a massive public effort to detect the presence of life in the universe.

> In 1983 one phase of the government-sponsored SETI (Search for Extraterrestrial Intelligence) project began at Harvard's Oak Ridge Observatory. It was designed originally to scan sixty-eight percent of the sky. The first multi-channel spectrum analyzer scanned 131,072 channels. This system was upgraded in 1985 to scan 8.4 million channels. NASA (National Aeronautics and Space Agency), with Stanford University, is developing a ten to fifteen million channel spectrum analyzer. This is planned to be operational in the 1990s.
>
> When the SETI project was operating at the Aimes Hershey Research Center in Mountain View, California, it was headed by Dr. John Billingham. Billingham said, "There is no doubt in our minds that intelligent life, far more advanced and complex than our own, is widespread in outer space. The United States wants to be the first nation to make contact. When we make contact it will be the biggest breakthrough in the history of mankind. These advanced civilizations could help us conquer problems like disease, pollution, food and energy shortages and natural disasters."[6]

Dr. Frank Drake, the pioneer of SETI, believes that 10,000 advanced civilizations exist just in our own Milky Way galaxy.[7] He relates that the true goal of the new one hundred million dollar upgrade is to "answer age-old philosophical questions about ourselves—Where did we come from? Are we unique? What does it mean to be human?" Dr. Drake believes that we will detect the broadcast signals of extraterrestrials by the year 2000—which, he says, "will profoundly change the world."

President Jimmy Carter, who himself claimed to be a witness of a UFO and who asserted a belief in the extraterrestrial theory of UFO origins, was a major supporter of the SETI program, and was responsible for much of its funding in the late 1970s. *UFO* magazine[8] quoted from President Carter's speech to NICAP (National Investigative Committee on Aerial

---

6 *UFO: End-Time Delusion,* p. 103.

7 *Omni,* "The First Word," October 1992, p. 6.

8 M. J. Ernst, "President Carter Launches New UFO Study to Probe Air Force Coverup:' April 1978, p. 30.

Phenomena) describing his sighting: "Carter said that his sighting took place around 7:15 p.m. one October night in 1969. The UFO was a sharply outlined, luminous globe 'about the same size as the moon, maybe a little smaller.'"

These attitudes and statements make us think that not everyone is "in" on the coverup, if one exists. Candidate Carter promised to reveal all government information on UFOs if elected. President Carter revealed nothing, but diverted many dollars to alien detection equipment. Apparently, no alien had yet been detected.

Invisible UFOs

It is assumed that many more landings occur than are observed or reported. As in the case of the Gulf Breeze sightings, it is possible that some UFOs possess the power to "cloak" themselves from human eyes, and come and go unnoticed. During Ed Walters' ninth sighting, he reports:

> Tonight was one of those "no hum" visits. It was 8:00 P.M., and, as I said, I was peacefully writing in my log when I noticed an easy shadowy movement slide slowly from the back of the high school gym along the dark grass field, closer and closer toward my back fence. The grassy field was dark green, and the textured shadow changed it to a fuzzy orange.
>
> It was clear to me now that the UFO could conceal itself from sight. The glow was there, but there was nothing above it. I had a clear view from the ground to sky and could see no craft.[9]

In an instant the UFO was suddenly there. As Frances Walters volunteered to telephone the police, the UFO disappeared. A moment later "the UFO popped back into view at about the same place it had left." One of the times a UFO seemed to disappear, a landing site was found the next day. The resulting ring of dead grass is the same discussed earlier.

The Tunguska Event

One of the greatest unsolved mysteries of the Earth, one that is also thought by some to have its roots in a UFO crash, is the immense Siberian catastrophe known as the Tunguska event. On June 30, 1908, a giant object

---

9 *The Gulf Breeze Sightings,* pp. 112-13.

entered the Earth's atmosphere and exploded just before reaching the ground. The explosion was so powerful that it blew full-grown trees to the ground for several miles surrounding the center of "ground zero," igniting a 1,200 square mile forest fire. The explosion is agreed to have had "nuclear" proportions, being estimated to have packed a 20 megaton explosive equivalent. The initial thought that comes to mind is that because there were no manmade explosive devices of that magnitude in 1908, the super-explosion must have been caused by a large meteorite. It was nearly twenty years before the first expedition teams entered the area, where they found the devastation of the trees and other vegetation, but found no craters sufficient to have been made by the meteor, and no radioactivity. Proponents of the "exploded UFO" theory point to the absence of a large crater as evidence that no meteorite could have caused the blast. Meteorite proponents point to a small cluster of ten craters, measuring from 30 to 160 feet across, having an average depth of ten feet. They feel that this is ample evidence that the giant meteor must have fragmented on entry into the Earth's atmosphere and exploded just before impact, creating the ten smaller craters. UFO proponents again indicate that the further absence of any signs of meteorite materials suggests that no meteorite ever impacted in the area. The smaller craters could more easily have been created by an exploding UFO they assert.

The Tunguska event will never be resolved definitively for the same reason no purported UFO landing site will be universally accepted as such; because as long as such impressions and discolorations can be explained by less exotic phenomena, we will accept the more ordinary explanations. The fact is that until a grounded UFO makes the network evening news, we will continue to demand more than just indentations and charred vegetation as proof of their existence.

Although many thousands of UFOs have been reported to have been seen on the ground, most of the sightings are either identical to visual contact with airborne craft, uneventful except for the sighting itself, or they fit into the following type of encounter—that of a very personal kind.

5
Extraterrestrial Visitors

The Nature Of UFO Occupants

As we enter the world of sightings of, and encounters with, alien beings, we quickly perceive that this is the "*Twilight Zone*" of UFOlogy. With the advent of direct contact with extraterrestrials, or whatever UFO occupants are, no clear answers present themselves concerning alien visitations to Earth—rather, the confusion intensifies. The answers we do begin to formulate leave us quite uncomfortable. Descriptions of the physical makeup of these extraterrestrial visitors vary widely. Accounts of such beings range from human looking, blonde-haired, blue-eyed men, to tiny, fur-clad gnomes with little in the way of facial features.

Some UFO proponents have attempted to categorize aliens into four or five basic groups. While some believe that the wide variety of alien forms is the result of our being visited by beings from different planets, others believe that we are visited by extraterrestrial humans, who in turn bring an entourage of "assistants"—robots to collect specimens, monsters to scare the earthling natives away, and lesser life forms to perform dangerous or mundane tasks. Other aspects of the natures and origins of the occupants of UFOs, *e.g.* spiritual, metaphysical, electromagnetic, and psychic are discussed below. Interestingly, proponents of human/alien theories often believe that much of the visitation is by humans or humanoids from this planet.[1] Two major Earth/UFO theories are dominant.

The Subterranean Civilizations Theory

The first theory is that human civilizations live in little-known subterranean regions of the Earth. These earthlings enjoy the benefits of monitoring us without being observed by us. They have all of our technology, plus that of ancient advanced civilizations, and some of their

1 Other paranormal theories exist, postulating that UFOs and their occupants are from parallel worlds that exist on other planes of existence.

own. There are people who profess to be former members of these civilizations. On a radio talk show heard by the author[2] the host conducted a telephone interview with a young (sounding) woman who calls herself Sharula Dux, who claims to be approximately 250 years old, and a former resident of an underground city, Telos, in the Mt. Shasta area of northern California. She said that this city possesses, among other things, a High Priest, for whom Sharula acts as a spiritual channeler to the surface world.

Sharula maintains that there exist more than 150 such subterranean cities around the world, which possess superior technology, and whose citizens live to extreme ages. The longevity is attributed to their toxin and radiation free environment, and certain healthy lifestyles that they lead. Interestingly, accounts from UFO contactees also tell us that UFO occupants live to ages in the 1,000-year range. It is fascinating that both Betty Luca (an abductee discussed later in the book) and Ed Walters, as well as others, indicate that they were taken under the water to subterranean complexes when abducted by UFOs. Significantly, these subterranean civilization theories have persisted for centuries.

The Returning Earthling Theory

The second theory explains that somewhere in Earth's history a human civilization possessing extraordinary technology left this planet and visits it from time to time. Accounts relate that these former earthlings have either moved to another planet,[3] or developed the ability of intergalactic travel at hyper-light speeds, causing their relatively short travels in space to translate into thousands of years to us. A subclass of this theory is that UFOs are piloted by our own descendants, who have mastered the science of time travel and have returned to help us.

---

2 KSL Radio, Salt Lake City, Utah, March 9, 1992.

3 Theories that we have descended from a race of humans from a fabled tenth planet in our solar system are proffered by some. The theory purports that our earthling forefathers returned to the home planet, but continue to visit us on Earth.

The Extraterrestrial Human Theory

Having touched upon the human/alien theories which hold that unknown earthlings are actually responsible for flying saucers, and reserving the "gnome and fairy" theories for later, the bulk of extraterrestrial personage encounters recount that human-like[4] persons are visiting our planet. Many accounts report that UFO pilots are nothing more than Earth-like humans, sometimes sporting larger ears or some other minor deviation, but essentially like us. The bulk of contemporary narratives, however, claim that "small greys" are manning the UFOs.

Small Greys

Most persons touting alien encounters describe alien beings that stand three-and-one-half to five feet tall, having grey colored, hairless skin, proportional arms and legs, diminutive mouths and noses, no ears, and very large dark or black "almond" shaped eyes set in a bulbous head with a pointy chin. These are the type of aliens described by Ed Walters, as well as most abductees, discussed below. These abductees generally only retrospectively "remember" the aliens, after "hypnotic retrieval" sessions with psychologists. Sometimes the small greys are accompanied, if not led, by taller greys or human-looking beings.

During the time the author lived in Italy 1978-1979, a wave of sightings occurred in that country. The effect was so alarming that even the Italian Parliament debated the subject. On July 4, 1978, Italian Navy personnel ascended Mt. Etna and sighted "three red pulsating UFOs, one of which landed." They described it as being a "domed disc about 12 meters [40 feet] in diameter with red and yellow body lights." They then observed "two tall golden-haired, white-robed beings accompanied by three or four shorter beings wearing helmets and spacesuits." During an inquiry, the military personnel said they "felt a compulsion" to climb the slopes. The phenomenon of tall humans accompanied by small humanoids is common

---

4 The term "human-like" is used broadly here. Humans on this Earth vary widely in size, color, characteristics, and appearance. If humans from other parts of the universe are actually visiting us, we could expect a reasonable amount of similarity or diversity without stretching the bounds of our understanding of the nature of humanity.

in reported observations of UFO occupants. Telepathic suggestion to rendezvous at a predetermined place is also a routine technique employed by UFO occupants. Ed Walters experienced at least two occasions when he saw the small greys up close, in full consciousness. On the first occasion, he was in bed when he heard the family dog bark in an unusual manner. He recounts:

> The pistol in one hand, the camera in the other, I walked over to the French doors that lead from the master bedroom out to a screened-in porch overlooking the pool.
>
> Cloth miniblinds covered the glass. I could tell it was still dark out. Only the faint glow from the school's security lights, across the field behind us, showed around the edges of the blinds.
>
> I felt for the draw cord and pulled it down quickly, leaning forward as the blinds came up. On the other side of the glass was a small creature. Big black eyes stared into mine. Just inches separated us. I screamed and fell backward onto the floor as my feet got crossed. My head and shoulders hit the closet door.
>
> The creature just stood there, staring in at me. It was maybe four feet tall. A dark, grayish-black, box-like thing hid most of its body. The "helmet" over its head had a clear insert that revealed its eyes, really big eyes that covered the top half of its head. It grasped a glowing silver rod in its right hand . . . .
>
> I still had the pistol in my hand, so I quickly raised it and pointed it at the creature. I wasn't going to shoot unless it tried to get through the door.
>
> The creature stared at me with eyes that showed no fear. Eyes that were calm. Eyes that were almost sad. Eyes that somehow seemed curious.

Ed Walters decided to attempt a capture of this small grey and ran out the back door only to be attacked by the blue beam. It is interesting that Walters recounts that this being wore shielding only in the front. As the creature walked away, its back was unshielded. The entire encounter was an apparent attempt to get Walters out into the open to abduct him with the blue beam. From this we surmise that the creature felt no need to protect itself in the back because it knew that Walters would be "beamed up" if he pursued the creature. Ed Walters saw the blue beam shoot down from the UFO just a few minutes after its failed attempt to capture him, and took a photograph of it. He speculates that the creature was beamed up in the blue beam.

The use of the blue beam as a transportation device gives rise to Walters's second conscious encounter with the UFO occupants. Ed Walters narrates that one evening while driving his pickup through a remote area to a construction site:

> I pressed the gas pedal and rounded a curve. Everything turned bright white. The hood reflected a brilliant flash. Some of the "light" came through the windshield and hit my arms.
>
> I yelled, "What the hell?"
>
> I really screamed, and at that moment, flash! Again I was being hit with a white beam. Almost like a flashcube going off inches from my eyes. It was extremely bright and left my eyes trying to refocus on the road. Within seconds, I realized I couldn't feel my arms. The sensation was the same as when your foot goes to sleep and you feel all those pin pricks . . . .
>
> Still accelerating down the road, I knew I was in trouble. The truck swerved from side to side and was almost out of control because I couldn't watch the road curves and watch my hands control the steering wheel at the same time.
>
> Only seconds had passed from the first flash of light. From overhead, and coming from behind, the UFO passed straight down the road in front of me . . . . I hit the brakes hard and came to a stop about 200 feet from the UFO . . . . By bracing my left arm on the steering wheel, I managed to shoot photo 19. I pulled out the film, preparing for another shot, when I noticed that the UFO was definitely moving closer.
>
> I panicked, afraid the white flash could hit me in the truck cab. Out of the truck and onto the ground I pushed myself in a gasping rush. I was scared, and my chest heaved as I hyperventilated. I was trying to crawl under the truck. The camera was slung on my left wrist, and I dragged the shotgun with my right hand.
>
> When I looked forward, down the road, the UFO wasn't there. I was halfway under the truck when it hit again. Flash? My legs stung and went numb from the knees down. I dug into the grass with my elbows and finally managed to make it to where my head was below the oil pan . . . .
>
> A blue beam flashed from the UFO to the road. Five times it shot down. Each blue beam deposited a creature on the road close to the UFO . . . .
>
> Finally all five began to move in lock-step toward me. Each one had a silver rod. They moved the rods up and down in their right hands as they marched down the middle of the road.

With this, Ed Walters decided to get into his pickup and flee, which he did, with great difficulty. These silver rods are apparently the same as that carried by the single small grey on the night it attempted to lure Walters out of the house. Apparently such a glowing silver rod is a technical device used by these beings as a stun gun, communicator, or some other environmental control device.

The Varied And Ephemeral Nature Of UFO Occupants

Descriptions of aliens, or whatever the beings are that occupy UFOs, range more widely than those offered above. "Moth-men" who fly without fluttering their wings have been reported around the world, as well as in the United States. Tiny two-foot-tall men with long white beards, or even giants, are commonly described. Many UFOlogists suggest that UFO occupants are ephemeral in nature, able to change their size, shape, and appearance at will. Some contactees report that the beings are self described as being able to change their appearance.

Kathie Davis describes a childhood experience in which, after seeing a flash of bright, white light and hearing a loud noise, she wandered to a "house" with the door open, although there was snow on the ground. Inside, she met a "little boy" who took her into his "playroom," where a small mechanical device surreptitiously cut her leg while the boy distracted her. Upon taking the blood sample she recounts that the boy metamorphosed into a small, large-headed, grey-skinned person.[5]

Shootout At Kelley

On a hot August evening, Billy Ray Taylor, a guest at the Sutton family farm in the Kelley area of southwestern Kentucky, went out to the well for a drink of water. He came running into the house in a panic, reporting that he had just seen a flying saucer fly over and descend into a gully near the house. The family teased him about his attempt to frighten them until the dog began barking nervously, and hid under the house. Billy Ray and Mr. Sutton looked outside and saw a very short being approaching the house. They describe the creature as having a large, round head with big, luminous eyes, and long arms with "talons" on the end. The two men went directly for some firepower—a 20-gauge shotgun and a .22 rifle. As the creature neared the house, they both began blasting. It appeared to take a hit, tumbling head over heels through the air, and disappearing. The men went outside to check for more creatures, finding one on the roof and one in the tree. They shot the one on the roof, which likewise somersaulted through the air, and then the one in the tree. This creature "floated" to the ground, escaping with a limping dash.

Mr. Sutton began around the comer of the house in search of more creatures, when he came upon one at the corner. He shot it at point-blank

5 *Intruders,* pp. 203-04.

range with the shotgun. He relates that he heard a loud hollow clink, like shooting a metal pail. The creature retreated, although it appeared to be uninjured. They surrounded the house, and the men shot at them through the windows, trying to calm the children between outbursts of fire. After three hours of this, the creatures finally disappeared. When it appeared that the creatures might be gone, the family dashed for the cars and drove to the nearest town for help.

The city and state police returned to the farm with the Suttons, and thoroughly investigated the incident. They brought along a photographer to make a detailed record of the investigation. The creatures were gone, but a luminous patch of liquid remained where one of the creatures had fallen after taking a blast. The police concluded their investigation early in the morning and returned to town to file their reports.

A short while later the creatures returned, again peering through the windows of the Sutton farm house. The men began anew their relentless firing of rifle blasts into the creatures through the doors and windows. After a couple of hours, the creatures departed for the last time. The only physical evidence of the second shootout was a house full of bullet holes. However, the police agencies involved were thoroughly convinced of the sincerity of Billy Ray Taylor and his hosts, the Suttons, and commented in their official reports that by the demeanor and attitude of the victims, the police believed their unusual story.[6]

Udo Wartena

Tim, a seminary instructor in the state of Oregon and an acquaintance of the author, learned from his father that an acquaintance of theirs had a close encounter many years before. After learning all that he could from his father about the incident, Tim interviewed Udo Wartena about the encounter while on a trip from Portland to Oakland, California, before Udo's death just a few years ago. Tim relates the pertinent portions of the interview from memory as follows:

> Tim: Do you remember talking to my father about your experience while mining for gold in the remote regions of Northern Idaho in the 1920s?

---

6 *The UFO Phenomenon,* Time-Life Books, Virginia, 1987 (*Mysteries of the Unknown* Series).

Udo: Yes. Your father was the first person I ever told about it.

Tim: That was some time in the late 1960s. Why did you wait so long to share your experience?

Udo: Who would believe what I experienced? Over the years I read some of the other accounts in the newspapers and magazines and realized that some of the same things happened to others. But some of them told of strange looking creatures. I doubted them, because the people on the UFO that I saw were just like us, and very nice chaps.

Tim: What actually happened? How did you know the ship had landed?

Udo: My mine shaft was nearly vertical and was over 30 feet deep in the ground. I had built up a sluice stream nearby to wash out the gold. I heard something like a big truck, or like a jet engine, but this was before the time when I had ever heard such a sound. I came up my shaft to look, and in the meadow was a saucershaped craft supported on legs.

A man was pulling down a hose to draw water from my stream. I did not feel the least bit of fear and I walked toward the man. He came and shook my hand, apologizing that they had not known that I was in the area. It was not their custom to interrupt or allow themselves to be seen. "Why not take water from the lake?" I asked. "This water is more pure and free from algae," he answered, and invited me to see his ship.

He was such a pleasant fellow. I asked his age—he was quite handsome and youthful, appearing to be middle-aged. His companion on board looked slightly older, but was also in very good health. They answered that one was about six hundred years old as we measured time, and the other was over nine hundred years old. They knew over five hundred languages and were learning ours, and improving upon them all the time.

They had come here "to monitor the progression and retrogression of our societies," he explained. They live among us from time to time.

I asked if they knew of Jesus Christ, and if they held the Priesthood. "We would like to speak of these things," he responded, "but are unable. We cannot interfere in any way."

"Where did you come from?" I asked. "We live on a distant planet," he replied, giving its name and pointing in its direction. The love, or comfort I felt in these men's presence was remarkable. I was invited to be examined with an "x-ray-like" machine, which when passed over me could record what impurities were in my system.

The ship was propelled with two rings or disks about three feet wide and a few inches thick, which circled the inside perimeter of the ship next to "battery/transformer-like" units all around the outer wall. Rods separated the disks (turned by motors), causing the disks to move in opposite directions. A force was generated which overcame the gravitational pull of the Earth, or any other planet they were on. They focused on a distant star and used its energy to draw them through space at speeds greater than the speed of light. My host specifically mentioned "skipping upon the light waves."

> They invited me to accompany them on their journey, but I declined—I was driving to Portland the next day for my wedding. I didn't want to leave them though.
>
> Upon leaving the ship and standing by a rock some distance away, I heard the same loud noise as before as the craft lifted, rotated with sort of a wobble, brought up the landing gear, and began to rise slowly. It went faster as it rose until it disappeared at great speed. An energy had permeated the area and I lost all my strength for some hours. I was unable to walk. When my strength finally returned I walked back to the base camp.
>
> When I arrived at camp no one else mentioned the craft—I never mentioned it to anyone of them. A young man came up missing from his claim, however. Nothing there had been disturbed—he was just gone without a trace. I was never sure if it had anything to do with my experience or not.

Udo returned annually to the mine for a few years following the incident, but never saw the craft or its occupants after that. He related that he always wished that he could have seen them again. Udo was so reluctant to speak of the incident that nearly five decades passed before he shared it with his wife—only after discussing it with his own church authorities. Tim describes Udo as a shy, pleasant, Scandinavian immigrant—a carpenter by trade. He trusts the word of this Christian man and reports that Udo related this account to his father in a "humble and matter-of-fact way, having a relationship of friendship and trust with him."

Although we can begin making a composite perceptual sketch of the nature of UFO occupants from the tidbits of information we gain from alleged observers, it is not until we commence an analysis of abduction reports that we gain any understanding of their purposes for visiting the Earth as claimed. The "space brother" preachings of the 1960s begin to pale in comparison as we compile the data that are streaming into investigation centers in connection with abductions, and the story they tell is quite unsettling.

6
Extraterrestrial Abductors

Missing Time

In the past few decades many people have come forward telling of perplexing experiences wherein they have seen a UFO, then have gone about their business without apparent interruption, only to find that an hour or two of time has escaped them, and no memory of the "missing time" can be recalled. The continuum of the passing of time seems completely uninterrupted to these witnesses, at least at first—yet there exist inescapable signs that something unremembered has occurred to them. This phenomenon is reported throughout the world. The Roper Organization poll cited earlier reflects that up to 13 percent of American Adults (over 24 million) have experienced such a missing time episode.

Judi Moudy

Judi Moudy is the legal assistant of Stan, the author's colleague in Italy. As we discussed the details of Stan's encounter on the telephone, he related that his assistant came to speak with him of her own encounter, as she was preparing his dictated account. He related that he was quite impressed with her account, as well as with her personally. He related that he has known Judi for a long while, trusts her completely, and knows that she is "incapable of lying." Judi tells of her encounter quite openly. Because it is so representative of missing time cases, we consider it first.

> My encounter occurred when I was in high school. It was August of 1977 and I was living in Del City, Oklahoma, a suburb of Oklahoma City. I was fifteen years old at the time and had two friends spending the night. Rhonda, Julie, and I decided to sleep in my parents' travel trailer which was parked in the back yard. Our home was located on the last street in a subdivision and was adjacent to a large alfalfa field. It was about 11:00 p.m. and *much* too hot and muggy to sleep. If you have ever been to Oklahoma City in August, you'll know what I mean. We fixed ourselves a coke and some popcorn and sat out in the front yard to talk. The adjacent town is Midwest City, where Tinker Air Force Base (home of the AWACS) is located, and my house is in the flight path of the east-west runway. Thus, I grew up with sonic booms, loud planes flying overhead and strange lights in the sky at night.

When Rhonda saw lights in the northern sky moving erratically, I told her it was probably just a plane coming in for a landing. All of a sudden the lights began moving toward us at a very high rate of speed. I had never seen a plane do this before. The lights kept coming toward us, getting bigger and brighter, until it was right over the field across the street. It was just breathtaking. Prior to this time, the crickets were chirping and dogs were barking. When the craft hovered over the field, all noises ceased.

The craft was cigar-shaped with lights around the top, bottom and center. The lights in the center actually looked like portholes as the light in this area was coming out of the holes inside the craft as if they were windows. The craft paused for about one minute, then slowly moved to the right to the far end of the field, then went straight down. Julie, the most adventurous of us three, took off running across the street to see where it landed. Rhonda and I followed, but when we got to the yard across the street we saw bright blue beams of light shooting through the alfalfa and coming straight at us. It was a very strange light—we could see the beginning of it, then it was as if someone turned off the source, and the beam separated itself and kept coming at us as if sections of pipe were being shot out. The beams were short at first, then kept getting longer and longer. Another strange thing was that there was a chain link fence at the corner of the yard across from us. When the beams became long enough to reach the fence, the light would *go over* the fence! Just past this fence was the concrete street . When the beam of light hit the concrete, it disintegrated into steam!

We kept watching the beams of light for some time. Suddenly we all became *very* sleepy. I looked at my watch and told Rhonda and Julie it was 12:30 and that we should get to bed because we had to get up at 6:00 a.m. for early morning marching band practice. The next thing I remember, we were back in my front yard walking toward the trailer. We were completely exhausted and went inside to go to bed . When we turned on the lights, we noticed that we each had dirt on our arms and legs. Earlier in the evening, Rhonda had given herself a manicure complete with bright red nail polish. She now had mud under her nails! I turned and looked at the clock; it was *3:00 a.m.* We couldn't understand what had happened. It seemed like I had just mentioned that it was 12:30 and that we should get to bed; we walked to the trailer and it was 3:00 a.m. I debated on waking up my dad, but he had been ill with ulcers and moody for some time. So we all went to sleep. When we awoke the next morning, Rhonda and Julie had terrible rashes on their arms and stomachs. I didn't have a rash, but I developed a *severe* headache which lasted for several days. I had never been bothered by headaches before, but I have suffered from migraines ever since that incident. We discovered that each of us had an identical inch-long cut on the outside of our index finger on our left hands.

We went outside to see if the craft was still there, but it was gone. We walked over to the field where it had been to see if there was any evidence of its presence. As we walked out through the field, the alfalfa was about neck high. Then we came upon an area where the alfalfa had been "burned" in the form of a triangle with three large circles at the ends [corners]. The alfalfa in the center looked similar to singed hair. We ran back to the house and told my parents, who were *very skeptical.* I finally convinced my dad to come to the field with us to look at the evidence. Julie's dad had already arrived and was now with us. After seeing the evidence our parents finally believed us.

Julie's dad had an interest in UFOs and knew of a researcher based in Norman, Oklahoma, named Hayden Hewes. Julie's dad called Mr. Hewes, and he came to my house that same afternoon to interview us. Mr. Hewes interviewed each of us separately with a tape recorder to be certain we each saw the same thing. Our stories concurred. Mr. Hewes took samples of the soil from the field, the alfalfa, and the dirt from under Rhonda's nails which he sent to a lab for analysis. Each sample was determined to have a high level of radioactivity.

My father was a civilian employee at Tinker Air Force Base at that time and called a friend of his who was working in the control tower that night to ask if anything strange had shown up on radar. The friend asked the reason for his inquiry and questioned him concerning what he had seen. My dad responded that his daughter had seen something strange in the sky and that he was just curious. The friend said he wasn't at liberty to give out any information, but the next day two gentlemen from the Air Force came to our house and interviewed us. They wouldn't give us any information either.

Reporters got wind of our story, did an interview, and printed it on the front page of the Daily Oklahoman newspaper. At the end of the article, Mr. Hewes requested anyone else who may have seen something that night to call his office. Twelve people called in, two of them police officers who wanted to remain anonymous.

Since my encounter, I have read almost everything I can get my hands on relating to UFO experiences. Unfortunately, most of the material seems fictionalized or sensationalized. Frankly, it scares me. It's hard for me to rationalize what may have happened to us—it just doesn't make intellectual sense. But I also don't know how to explain the loss of time and the identical scars.[1]

Judi's experience is much like those of thousands of others. Several questions leap out at us and demand explanation: what happened during the missing two and one half hours? how did the girls become so dirty? and, what are the identical cuts from? No explanation can be had absent the restoration of the girls' memories. However, others with similar experiences have remembered, or subsequently recalled what happened to them.

Between the time that Judi sent the copy of her account, and the time we discussed it on the telephone (about a week), the television movie *Intruders* aired. To take advantage of the market hype generated by the network

1 Letter to Author, dated May 12, 1992.

advertising for *Intruders*, much of the week's television programming included UFO-oriented programs. Especially impressive were the accounts coming out of Belgium and other areas concerning triangular UFOs. Thousands of official reports have been made to the Belgian government since 1987, including many from government employees—police, etc. These craft appear to be "cigar-shaped" when seen from the side, with a row of lights or lighted windows around the perimeter, as reported by Judi. When seen from below, however, witnesses report that the craft are actually triangular, having rounded, convex edges all around. The design is much the same as if one bent a pipe into a triangle and covered the inside hollow section with a covering, top and bottom.

The UFOs reportedly had white glowing lights on the underside of the three corners. It is interesting how precisely similar these reports are with Judi's 1977 report—cigar-shaped from the side, yet leaving a singed triangular impression in the alfalfa field with distinctly round markings at the three corners. This similarity did not escape Judi either.

The Betty And Barney Hill Abduction

One of the most famous UFO abduction cases, and the earliest to be widely reported, is that of Betty and Barney Hill. Their experience, as told in John Fuller's *The Interrupted Journey* (1966) and a series of *Look* magazine articles, typifies many contemporary abduction stories. The Hills were reportedly driving late at night in the fall of 1961. As they made their tedious journey through the White Mountains from Montreal to their home in Exeter, New Hampshire, they spotted a bright light in the night sky that began to follow them. Barney stopped the car to get a better look at the object, when they noticed that it was a craft with windows, through which the Hills could see occupants. When the Hills saw this, they jumped back into their car and drove home, arriving two hours later than they thought the journey should have taken.

Within a week or so the Hills started suffering from recurring nightmares wherein they were forcibly abducted and taken aboard a UFO. Their captors were small greys, with the exception of a taller leader and "doctor." This doctor is so described because his function was to perform certain physical examinations on the Hills. Betty was examined almost exclusively as to her reproductive organs. She describes under hypnosis an examination technique that was completely unfamiliar to her, or anyone else at the time:

> The examiner has a long needle . . . . He said he wants to put it in my navel, it's just a simple test . . . . And I'm telling him, "It's hurting, it's hurting, take it out." And the leader comes over and he puts his hand, rubs his hand in front of my eyes, and he says it will be all right. I won't feel it . . . He said it was a pregnancy test. I said, "That was no pregnancy test here."[2]

It was many years later that our own technology developed the correlating procedure of laparoscopy to perform *in vitro* examinations. In fact, laparoscopy was later applied to treat infertility by placing sperm and oocytes directly into an infertile woman's fallopian tubes for *in vitro* fertilization.

Barney Hill was similarly subjected to examination of his genitals, reporting that a circular instrument had been placed over the groin area, later resulting in the appearance of a circle of warts at the site of the instrument.

Other than their recurring dreams, the Hills had no conscious memory of the abduction. A few years later they sought psychiatric assistance, not understanding the reasons for the underlying stress they were experiencing in their lives and marriage. It was while under psychiatric hypnosis that the Hills independently recalled their harrowing experience of abduction and physical examination. While under hypnosis Betty was able to recall and draw a star chart she had observed on the UFO. Some feel that the chart she drew is extraordinary because it matches perfectly with far away star systems. Debunkers feel that the chart is insignificant because of the high probability of matching any given star pattern with a true charting, given the large number of stars in our universe.

---

2 Raymond E. Fowler, *The Watchers: The Secret Design Behind UFO Abduction,* Bantam Books, New York (1990). Raymond Fowler cites the Betty Hill examination to compare with that of the subject of his book, Betty Andreasson Luca: "Oh! And he's going to put that in my navel! Ohhhh! Feel's like he's going around my stuff inside—feeling it—with that needle." Like Betty Hill, Betty Luca also received the benefit of the abductor putting his hand on her head to relieve her pain.

Onboard Medical Examinations

The disturbing pattern of UFO abductions that appear to be inundating investigators are those in which abductees report being taken aboard UFOs and given medical examinations. Often, as with Betty and Barney Hill, these reported examinations are centered in human reproductivity. Stories abound that human semen and egg samples are collected during these abductions. Others, men curiously, claim that they are forced or enticed to breed with a female alien or human/alien hybrid. Generally, however, sperm samples are taken mechanically. Women further claim to be artificially impregnated with alien semen.

Many reports are surfacing that offspring of these human/alien reproductive experiments exist. Accounts are becoming more common that women who were in their first trimester of pregnancy are suddenly finding themselves no longer pregnant. Medical experts reportedly confirm that a *bona fide* pregnancy existed, and was then non-existent, leaving no trace that a pregnancy ever existed or that an abortion occurred. As much as this all sounds like fodder for selling supermarket tabloids, many UFOlogists are buying into the phenomenon. They report that upon the application of hypnotic retrieval to such a person, memories of multiple UFO abductions surface. In the first instance, generally in childhood, abducted youth are subjected to the taking of tissue samples. Later, as young adults, those who pass genetic muster are abducted for the taking of sperm[3] and ova. Sometimes, the abducted women are impregnated, generally artificially, and later the developing fetus is removed from the womb and a special therapy is applied to remove signs of pregnancy. Budd Hopkins documents in *Intruders* how multigenerational abductions have created experimentational bloodlines, evidently of high value to the abductors.

---

3 Two of the men that Budd Hopkins worked with expressed that they had received vasectomies, due to an overwhelming, innate drive to do so. The one reports that he was later abducted for the purpose of mating with a hybrid woman. (*Intruders,* pp. 138-40) He recounts (through tears of humiliation) that after the perfunctory sexual encounter (unemotional for her, traumatic for him), he was nearly thrown out of the UFO for his misdeed. He received an unequivocal telepathic message that his captors were not amused.

Antonio Villas-Boas

Before the medical examination/reproductive experimentation aspects of the UFO abduction phenomenon were widely reported, a young Brazilian student, Antonio Villas-Boas sheepishly reported a "rape" in 1957. Antonio was helping out on his father's farm, plowing a field by night, when an egg-shaped UFO landed within fifty feet of his tractor. The tractor turned off by itself as three or four short uniform-and-helmet-clad creatures exited the craft and came toward him. Antonio attempted to escape, but the creatures dragged him into the UFO. They took blood samples and removed Antonio's clothing. After rubbing his body with a clear liquid, the creatures put Antonio into a room by himself aboard the UFO. The room soon began to fill with a smoke, which made Antonio nauseous. A small, blonde, naked female "woman" then entered the room, and after some prodding, was able to lower Antonio's defenses, and sexual intercourse ensued—twice. Antonio claimed that the liquid rubbed on him must have lowered his resistance to her promptings. Following intercourse, Antonio recounts that the woman pointed to her abdomen, and then to the stars. Investigators felt that her mute message indicated that a child would be born from their encounter.

After dressing and receiving a quick tour of the UFO, Antonio was ejected from the craft and left to himself. His story seemed more fantasy than fact at the time, except for a few items of latent evidence. Antonio became quite nauseous over the next few days, during which time strange sores developed over his body. He was later examined by a local physician who was unable to diagnose or cure his symptoms. Some experts have conjectured that his symptoms were very much like radiation poisoning. This possibility is echoed by the reports of others who are told by their abductors that their short stays on the UFOs are due to the fact that they will become ill if they stay too long. In any case, no matter how fantastic Antonio's tale appeared in 1957, the fact that many similar reports have surfaced since then, coupled with Antonio's reputation for honesty, have made his account one of the most accepted UFO "rape" cases in the abduction literature.[4]

---

4 Jerome Clark, *UFOs in the 1980s,* Apogee Books (1990), p. 10; *The UFO Phenomenon,* Time-Life Books, p. 10.

Human Accomplices

Kathie Davis reported an eerie encounter at a mountain cabin that she and others had with three young men who were plainly unconventional. This encounter was accompanied by a UFO sighting,[5] and missing time for all involved, including the adult cabin owners. They all relate that the "leader" of the three did all of the talking, and was very interested in Kathie (not a good-looking young woman). The behavior of the three was quite suspect. For instance, they mentioned that they were together because they played in a band. When asked what kind of music the band played, they first asked what kind of music those present preferred, and then responded that it was the kind of music they played. This sort of seeking out appropriate responses to questions was prevalent during the encounter.

As Budd Hopkins questioned those present about the incident, they all reported that they had absolutely no recollection regarding the appearance of the two taller, quiet young men. They all, however, remembered every detail about the leader. After individual descriptions of the leader, Hopkins remarked to each that his or her characterization sounded like a description of Kathie, to which they each agreed there did exist a strong family resemblance. In light of the leader's unusual interest in Kathie's life and background, and the multigenerational genetic and reproductive research apparently occurring in the "Davis" family, Hopkins's implication was that this young man was an unknown brother of Kathie's, retained by the aliens after conception, performing reconnaissance missions for the UFO occupants. The concept of human agents of UFO occupants is not at all new.[6]

The Nursery

As an epilogue to the fetal removal phenomenon, some hypnotic retrieval attempts have unearthed memories of yet another abduction during which

---

5 This UFO sighting is noteworthy in that Kathie Davis describes the UFO in terms of seeing "four lights descending, spinning like pinwheels." Other than reports of the main body of a UFO slowly turning on its axis, few accounts mention "spinning." This "spinning like pinwheels" description is the closest account the author has located to the "Wheels of Ezekiel" UFO theory proffered by those of the "Biblical UFOs" bent.

6 *See generally*, Jim and Coral Lorenzea, *Abducted,* Berkley Publishing Corporation; e. g., "The Walton Affair," pp. 80-86.

the donor of the ova or fetus, or in some cases, the contributing father, is given the hybrid child to hold and cuddle. After the "bonding" session is completed, the child is removed from the mother or father who is then returned to the bedroom or other place of abduction, without any memory of the abduction or the child. Theories attempting to explain all of such reproductive experimentation by aliens speculate that genetic manipulation is occurring, either to enrich alien gene pools or to regulate human evolution. A third possibility is that a hybrid race is being created for an unknown purpose. Of course, those of the Erich von Daniken bent claim that man was created by this very process,[7] and that the course of his evolution has always been determined by the genetic interference of our alien "fathers."

These hybrid babies are generally described as tiny and thin, often lying in small metal containers in the UFOs, having thin, greyish skin. The human women are very attracted to the babies, finding them almost irresistible—with hypnotic eyes. It is while "falling into" the babies' dark eyes that the women develop a sense that the babies are "wise," or "omniscient." A type of communication occurs between the women and the babies, during which the women become captivated by the great wisdom and intelligence of the babies, feeling a close kinship to them. Kathie Davis described her feelings about her little hybrid, Andrew, as she looked into his eyes as, "It's like the whole world was in this little baby's eyes. It was like, God, he knew, *he knew* what I felt. He just knew. I can't describe it. It was so intense, so euphoric or something. I was so excited, so up."[8]

During such an abduction experience of a psychotherapist called "Lucille," she gained information that offers an explanation of why these reproductive experiments are occurring. She "sensed" from her telepathic communication with the aliens that their society was "millions of years old, of outstanding technology and intellect but not much individuality or warmth." She wrote "the society was dying, that children were being born and living to a certain age, perhaps preadolescence, and then dying." She felt that there was "a desperate need to survive, to continue their race."

---

7 Von Daniken theorizes that humans were created by altering the genes of Earth apes with genetic information of extraterrestrials. This explains the missing link in our evolutionary process according to von Daniken.

8 *Intruders,* p. 184.

Lucille explained, "It is a culture without touching, feeling, nurturing . . . basically intellectual. Something has gone wrong genetically. Whatever their bodies are now, they have evolved from something else. My impression is that they wanted to somehow share their history and achievements and their present difficulties in survival. But I really don't know what they are looking for." Lucille was shown a series of "holographic" images, remarking, "I saw a child about four feet tall, gray, totally their race, waving its arms . . . it was in pain and dying. I was told that this is what is happening now."[9]

The "Aliens" Have Something To Learn From Us

An underlying theme of many of these "nursery" abductions is a sense that the aliens are sorely lacking in "warmth" skills. Kathie Davis claimed that she was abducted and shown a small hybrid female child, and was told it was hers. She described her strong feelings of love for this "beautiful" little half-human girl. She explained that the female aliens, distinguishable only by their emotional "feel," were interested in learning how she held, cuddled, and loved the child. On a subsequent abduction, Kathie's little girl, about four years old, joined in watching and learning as Kathie was told that an additional eight children had been produced from her removed ova. She was then allowed to hold the youngest of her many hybrid children. She recalled under hypnosis:

> They want to watch me . . . hold this . . . baby. They want . . . to feel how I love it. I shouldn't worry, 'cause she'll [the first female child] take care of it. I have something they can't give.
>
> [When asked "what?" she replied;]
>
> Something . . . to do with touch, and the human part . . . and they don't understand, but they'll learn. And they said I could name them. I would choose . . . .

Kathie noted that this undersized hybrid baby "was all pale . . . he looked dead, but he wasn't." When she began cuddling and kissing him, he suddenly gained strength and vitality.[10]

Lucille reports that in her encounter, she noted the difficulties of the alien race:

---

9 See *Intruders,* pp. 187-91, for a full narrative of Lucille's abduction experience.

10 *Ibid.*, p. 184, see also, *UFOs in the 1980s,* p. 10.

> We spoke about the lack of touching. I told them that some animals here can die within a day of birth if they are not licked and touched by their mothers or other loving caretakers, since that affects their perceptions of their bodily functions as well as of themselves. Strange as it may seem, I suggested their interpreting Ashley Montague's book *Touching.*[11]

Erased Memories

Most of these occurrences of abduction, and their often attendant reproductive system experimentation, are not "remembered" initially by the abductees. As in the case of the Hills, unexplained periods of "missing time" or recurring dreams or other subconscious manifestations of abduction and experimentation plague the person until psychological assistance is sought. Hypnotic retrieval of these repressed memories often recovers the memory of the missing time or forgotten period and the details of the abduction. The conscious memories of the abduction appear to be purposely "erased" by the abductors. Some abductees claim to have been told by the aliens that because no one would believe them anyway, it would be better just to have their memories erased. Other abductees have seemingly received forced amnesia through less beneficent means. Laura, Kathie Davis's sister, reports that hypnosis to help her lose weight had the opposite effect, making her eat everything in sight. When she called her hypnotist, upon hearing his voice she became homicidally violent. When taken to another hypnotist, it took several sessions to get her past an implanted posthypnotic suggestion that she would die if she were hypnotized or if she tried to remember her abduction experience.[12]

Hypnotic Retrieval

The practice of hypnotic retrieval is controversial because debunkers claim that those who perform the hypnosis "suggest" or "plant" ideas of abduction in the hypnotised person, even if it is done unintentionally. Debunkers fear that hypnotists inadvertently "lead" the hypnotized person, or "indicate" appropriate responses to questioning without intending to do so, thereby creating abduction memories. Budd Hopkins, presumably the

---

11 *Ibid,* p. 191

12 *Intruders,* p. 12.

foremost (or most experienced) UFO hypnotic retrieval expert, claims that he has certain "control points" by which he discerns abduction reports hypnotically retrieved. Hopkins claims that there exist certain consistent "details" in most genuine "retrieved" abduction cases that he withholds from the public. Because only he and his colleagues know what these details are, abductee wannabees have no way of knowing how to fake a hypnotic session with him. Some of these control points are specific examination and surgical techniques used by the abductors and specific hieroglyphs and other symbols seen while aboard the UFOs. Although this control point technique may satisfy Budd Hopkins, it does little to alleviate debunkers' fears of inadvertent memory implanting by the hypnotists.[13]

The Ed Walters Abductions

It may be no surprise that Ed Walters was also abducted by the UFOs that plagued his life for five months. He too had no conscious memory of being abducted, but was able to recall some details of abduction through hypnotic retrieval. He begins with a narration of what he consciously remembered about the abduction:

> In a fraction of a second my eyes caught the glow of the UFO about thirty-five feet above me. I flinched, and my right hand squeezed the shutter button, resulting in photo 39. Then, with the UFO just above a small oak tree, my eyes went completely white, just as if a flashcube had gone off in my brain. I could not tell if the whiteness was all around me, like a floodlight, or just within my head. Instantly I felt nothing, no sensations from my body, only a vague sense of falling.
>
> The next instant I was lifting my face and chest up off the sand at the edge of the water. My head pounded, and as I tried to stand, I stumbled in dizziness. Disoriented, I crawled up the beach twenty feet to the bench and sat there with my head in my hands.
>
> A smell from my hands was making me nauseated. Then it dawned on me. How could I fall to the ground while standing at the [camera] and get up a second later twenty feet away? I checked my watch and couldn't believe the time. It was 2:25 A.M. What had happened to the hour and fifteen minutes[14] between taking the photographs and finding myself on the beach?[15]

---

13 See generally, *UFOs in the 1980s,* pp. 1-13.

14 Most reported UFO abductions are of short duration, lasting generally one to eight hours—usually closer to two. Although most of the time the UFO occupants are said to have no more interest in the abductees than to hurriedly complete their predetermined examinations and medical procedures, Kathie Davis reported that on more than one occasion she was told she had to leave because a delay would result in her illness.

15 *The Gulf Breeze Sightings,* p. 263.

With the aid of regressive hypnosis Ed Walters claims that he is discovering that he was abducted and taken aboard the UFO on that occasion, as well as on other occasions during his life. He recounts three weird (terrifying, occult-like) experiences that occurred at eight-year intervals in his life, wherein missing time was present. Walters again asserts that through hypnotic retrieval he has learned that he was abducted on these occasions also.

Although Ed Walters fails to offer details of why he was abducted, he believes that the abduction on the Gulf Breeze beach will be his last. He offers in support of this belief the following facts which occurred upon his waking the following morning.

> In the bathroom I ran a comb through my hair and felt a bump at the back of my head very close to the center of my neck. It felt bruised, and I went to the mirror to try to see what I could feel. When I looked into the mirror I couldn't see the lump, but as I turned around, I immediately saw more than I expected to.
>
> A large bruise, with a red dot in the center, was prominent between my eyes right at the bridge of my nose. Two more similar red marks were centered on my temples, each surrounded by a bruise. I was shocked.[16]

Implanted Monitors

Researchers of the Ed Walters case believe it probable from their observations and Walters's reports that he was abducted as a child, at which time he received certain tiny implants around his head. The abductors could monitor Walters as well as communicate with him through the implants. Walters was monitored for an unspecified purpose by the communications implants and picked up at regular intervals for examination. This implantation and periodic pickup scenario is becoming quite common in UFOlogy.

All of this sounds uncomfortably like our own methods of tagging and tracking animals in the wild. In fact, some investigators are surprised at the high number of reported implants that appear to be occurring. Budd Hopkins claims in *Intruders* that the implanting of small "BB-like" objects in the abductee's inner ear or nose is becoming a disturbingly common complaint. The most common method appears to be the nose implant.

---

16 *Ibid.*, p. 264.

Hopkins recounts that some of his abductees "have recalled a thin probe of some sort with a tiny ball on the end having been inserted in the nostril, and they feel pain when the probe apparently breaks through at the top of the nasal cavity."[17] Abductees similarly report such probes being placed into their noses or ears, and seeing no tiny ball until the probe is withdrawn. This, of course, would be at the time of removing the previously implanted device. Regardless of the point of penetration or final lodging, these tiny balls are evidently placed in close proximity to the subject's brain. A handful of children of abduction families in the Copley Woods area (Indianapolis) were discovered by Budd Hopkins to have suffered terrible nosebleeds during the night. One mother remarked to Hopkins that a doctor at the local hospital said her child had probably put a pencil or a similar object up his nose, and punctured the membrane, as evidenced by a small wound there. He then observed how it was peculiar that so many children had been brought into the hospital recently for identical wounds.[18]

Dreams Or Memories?

Ed Walters was only beginning his regressive hypnosis therapy when he published his book, so few details of his boarding the UFOs are reported therein. However, one initial indicator of suppressed memories mentioned in the book is a vivid, frequent dream, which later appeared to be a subconscious memory surfacing. In reciting the details of the dream, Walters reported:

> The dream would begin with me rising high in the sky and looking over a coastline. I could see the sandy beach with waves breaking on the shore. Sometimes I would recognize the beach but most often not. Then I would quickly descend and pass beneath the water into the ocean. I would gasp for air, in fear of drowning, but as I went deeper and deeper, I realized I was inside a container with a large diamond-shaped window.[19] Through the window I could see the water and fish. Shortly thereafter I saw a lot of bubbles passing in front of the window, followed by rising sand, which soon completely covered the glass. That's all I remember of it.[20]

---

17 *Intruders,* p. 44.

18 *Ibid.*

19 The photographs of the Gulf Breeze UFOs reveal diamond-shaped portals surrounding the UFOs.

20 *Ibid.*, p. 161.

Ed Walters writes that during some of his encounters with the UFOs, as he would attempt to avoid abduction, the voices he heard commanding him to step out into the clear and not resist would taunt him with clues about what was occurring in his life. Upon one failed abduction attempt the UFO voice said, "Zehaas . . . in sleep you know . . . we are here for you." After cursing at the UFO, Walters again heard, "Zehaas . . . sleep and know."[21] Interestingly, the before-cited Roper Organization poll found that a full 10 percent of the adult American population has experienced similar, vivid dreams.

Many abductees remember portions of their ordeals as bad dreams. Kathie Davis still refers to her experiences as her dreams. Mary M., thought for many years that she was only experiencing nightmares. It was when she began reading of the accounts of other abductees that she discovered that her nightmares were exactly like those of hundreds of others. She, like many so afflicted, has come to believe that more than "dreaming" is occurring in her life. Of course, for many, the answer could be nightmares—or phenomena not based in reality, or at least not the realms of the normal physical world.[22]

The Bedroom Encounter

In Walters's case, as well as the majority, a number of abductions or other encounters occur while the subject is cozy in his or her own bed. These "bedroom encounters" are classic, and if the subject remembers anything at all, it is generally the very end of the encounter, having a foggy recollection of a small grey exiting the bedroom. Again, the Roper Organization poll reports that 18 percent of American adults have had this kind of terrifying bedroom experience.

Kathie Davis reports a few bedroom encounters, one of which is more than classic. She remembers it as more of a dream than reality, although completely vivid, as is common. It was still dark, and Kathie found herself sitting up in bed, awake, facing two small greys standing next to her. One

---

21 *Ibid.* p. 166; see also, p. 241.

22 See generally, *UFOs in the 1980s,* pp. 1-13.

held a black box with a glowing red light on top, and moved forward to hand it to Kathie. Again, as he moved, the second one duplicated his movements precisely, although his hands were empty. Kathie was terrified, and the experience seemed too real to be a dream. She says they spoke to her like a child, and called her by name. She narrates:

> He handed me the box. I said, "Can I have it?" He said, "No. Hold it. Look at it." I did. Then he took it from me gently after a minute. I said, "What is it? What's it for?" He said, "Look at me." Then I thought, "Do I have to?" but I did. He said, "When the time is right you will see it again, you will remember and you'll know how to use it." I said, "O.K."[23]

Upon waking, Kathie told her husband, who had slept through the visit,[24] what she had "dreamed." No subsequent information ever revealed the meaning of the cryptic statements made by the visitors regarding the box,—not yet, anyway.

Underwater Joyrides

Through hypnotic retrieval Ed Walters came to believe that his dream of flying through the air and into the water was a suppressed memory of an abduction and "joyride." It is interesting that in interviews with the investigators of the Gulf Breeze sightings, this dream phenomenon was largely ignored. If the abduction and "joyride" actually did occur, it is significant that Walters describes that the UFO went into the water and beneath the surface of the ocean floor into the sand.[25] One of Walters's previous missing-time episodes occurred while he was in a canoe on the coastline. On that occasion he saw a green glow under the surface of the water and air bubbles just off the bow of his canoe. The next thing he knew, six hours had passed and his sandwich was stale. Significantly, many abductees relate experiences of being taken beneath the water to underground or undersea complexes. Even if the explanation for UFOs lies

---

23 *Intruders,* p. 14.

24 It is common in bedroom encounters for the spouse to sleep through the incident, no matter how much conversation occurs, or how much the victim attempts to wake the spouse.

25 *The Gulf Breeze Sightings,* p. 161.

in a realm other than reality, it is fascinating that this sort of congruency exists between the accounts of so many otherwise unrelated encounters and reports.

Many other observers in the Gulf Breeze, Florida, area reported seeing UFOs flying into the water. In fact, UFOs flying in and out of the water has been a fairly common phenomenon in UFOlogy. If these UFOs can be traced to the ocean floors, and even beneath the sandy bottoms, researchers would do well to expand their search to those areas as possible Earth bases of extraterrestrials, or places of origin for terrestrial UFOs.

Travis Walton

In the small town of Snowflake, Arizona, lives a community of tightknit Christians. One of them, Travis Walton, has one of the most celebrated, credible tales of abduction in UFO lore. Most UFO books and "Unsolved Mystery" types of television programs include Travis's experience in their roster of top encounters. The motion picture *Fire in the Sky* is based on his ordeal.

One evening Travis and a group of his friends were travelling home in their pickup through the woods after a day at work. As they talked and laughed, they suddenly saw a bright light through the trees. They thought it might be an unreported forest fire so they drove near it. As they approached, they saw that it was no fire. Travis was nearest the door so he got out and had a look. He describes how he saw a bright craft hovering off the ground. It was not obscured nor hazy, but had a "clear and distinct" form—Travis says it was so close he could have thrown a rock at it.

The thought occurred to Travis, foolishly in retrospect as he narrates, that the thing "might take off" in a moment, so he moved in to get a "closer look." As he stepped forward, there suddenly was a bright flash which knocked Travis to the ground. Travis's buddies in the pickup were already of the opinion that Travis should not have left them in the truck to wander near the strange craft. Upon seeing these otherworldly hostilities and Travis's apparently dead body lying motionless on the ground, they sped off in a panic.

Travis's best friend and future brother-in-law, Mike, was at the wheel. Within a quarter of a mile, Mike began to regain his thinking processes, and remembered that Travis was lying alone in the dirt back there. He stopped the truck and an argument about returning ensued. Soon, all were thinking clearly and they returned to the scene. Travis was nowhere to be found, however. No sign of the strange craft could be detected either.

The frightened men all held hands as they warily searched the surrounding forest for Travis. They could find no sign of him. They decided to get help from the authorities, and raced to the sheriff's office in town. The sheriff was summoned to the scene, and after a brief search, decided to continue their man hunt in the morning. Already, however, his suspicions of foul play were growing—tales of UFO abduction were new to him, too. Travis's mother was notified of his disappearance.

The next morning the sheriff's office and Travis's friends conducted a thorough search of the area, finding no clues. The men recounted how they had seen the UFO and how Travis had approached it, only to be knocked to the ground by a brilliant flash. The sheriff developed serious doubts about the UFO story, and began investigating Travis's disappearance as a homicide. Mike was dating Travis's sister at the time, and the sheriff suspected that Travis had objected, Mike killed him in a dispute, and the other men helped bury Travis's body.

The group of men were questioned extensively over the next few days, and a polygraph expert was called in to check their stories. The polygraph expert had never before had so many prime witnesses to the same event. After subjecting all of the men to examinations, he concluded that they were absolutely telling the truth. So—where was Travis?

Five days after his disappearance, Travis Walton walked out of the woods. He narrates what happened in the interim:

> When I felt the numbing shock, I blacked out. And the next thing I knew, I regained consciousness—not quickly, sort of gradually. My head wasn't real clear, I was in a lot of pain. I was laying on my back—I didn't know where I was. I remembered what had happened in the woods. As I was regaining consciousness I was trying to figure out where I was and what was going on. I thought maybe I was in a hospital or something—I had been hurt.

As Travis looked around him, he discovered he was lying on a table in what looked like an examination room, but that he was in no hospital—not one that accepts Blue Cross anyway. He saw two small greys at the end of the table and jumped off in a panic. He backed away from them as they stepped towards him. Travis continues:

> When I was standing in front of those things that were coming towards me—and they stopped there, and they stood there looking at me—these huge eyes, just seemed to look right through me. I didn't get any impression of emotion, it was very detached—sort of, a just observing sort of thing, but it seemed like they could see everything I was thinking of doing. It was very disturbing to me to feel so . . . exposed. And these huge eyes looked at me and . . . when they'd blink, and on an eye that big the eyelid just slid down and opened like a window opening and shutting—and it just had the strangest sort of feeling, and I just couldn't, I couldn't bear their gaze.

The two small greys walked toward Travis as he threatened them with a glass rod he found and shouted out "Get back! Get back!" They walked right past him and out of the room. Travis took off, running wildly through the slippery, illuminated maze of corridors.

Travis finally arrived in a room with a high-tech chair in the middle. He was alone, so he went over to look at the chair. He recounts:

> There was a lever there, and when I moved that, the star pattern [that had appeared in front of him] appeared to move. That kind of disoriented me for a minute, because I felt like I was moving, kind of, for a second, because this was—to have everything—I had, by that time surmised, that I was in some sort of craft—I connected it to what had happened before, and figured I might crash this thing or something.

As Travis looked around the room and wondered what might happen next, a tall, young-looking human man with long hair entered the room. Travis explains:

> This person was not like the humanoid creatures that I had seen earlier. This looked like a human being. It looked like a man, in a blue uniform. I went up to him, thinking that I was being rescued, that I was being saved—that this was a person. I started asking all kinds of questions "Where am I? Where are we? And who were those things that I saw? Talk to me!"

The young human silently put his hand on Travis's shoulder, and led him out of the room into the corridor, and through another doorway.

> When that door opened there was an inrushing of air, and it felt fresher and cooler than where I'd been. It must have been, like—the air I was in was real heavy, moist, stifling.
>
> He just pulled me quickly on—went through some doors, down a hallway to another room.

In the other room, Travis and the young human were met by two more humans, another young man and a female. They maneuvered Travis, who was protesting and fighting, into an examination chair, something like a dental chair, and applied a "gas mask" to his face. He lost consciousness at that point, and next found himself waking in the forest, five days after his abduction. Like his friends, Travis was also subjected to a lie detector test concerning his disappearance and abduction, which he passed. However, some accounts relate that Travis's first test was inconclusive, but that subsequent polygraphs confirmed his belief in the reported experiences.

Travis has suffered a great deal of "frustration and pain" as a result of his unusual encounter—not resulting so much from the trauma inflicted by his abductors, but from that inflicted by those who accuse him of deceit. Everyone he meets fails to see him as a "person" now, he complains, but they see him as "that man," the "abductee." Travis laments, "Every contact I have with people is colored by, and filtered through the distorting lens of something that just happened to me—fifteen years ago." This is the reason that most people do not report their encounters.

7
Strange Encounters

With so many reports of UFO sightings and close encounters with "alien" beings (estimated by many to be in the millions), it is no wonder that retail book stores and public libraries offer accounts of thousands of bizarre brushes with the phenomenon. Two notable encounters are just beginning to receive publicity, and are included herein because they are instructive in our search to discover the source of the beings. As elsewhere herein, the accounts are included without commentary and at full face value to provide information about the state of the phenomenon.

The Mojave Triangle

Steve and Dawn Hess lived a quiet, normal life in southern California, with their two-year-old, Steve Jr., and newborn girl, Bethany. Steve was a devoted husband and father, and a hard-working supervisor of a construction company. Dawn was a loving wife and mother, choosing to stay at home with the baby. The Hess's were well-educated, conservative people, Dawn springing from very conservative Christian roots.

Steve was the rugged type, a hometown football hero and avid deer hunter. Although Dawn was athletic, she had little interest in hunting, and no interest in killing trophy-sized bucks in the Mojave Desert. However, in late October of 1989, Steve was able to persuade Dawn to go hunting with him for a few days by throwing in some sightseeing and entertainment along the way. She succumbed, and the children were left with their grandparents for the three-day hunting trip.

The first day of their excursion was spent driving and sightseeing, as promised, and they ended up pulling off the road for the first night. There, they slept in the back of the pickup, in the comfortable bed with all of the necessities, protected by the canopy shell. The next day was spent hunting and sightseeing, ambling toward Steve's favorite hunting spot. When they arrived at the BLM campsite, which had never been full in Steve's twenty-year recollection, they were astounded to learn that there was no room for them. Steve drove them out to a campsite in the desert where he and his family had camped many times, where they set up camp and cooked dinner.

Dawn was a little apprehensive about camping alone in the middle of the desert—more concerned about motorcycle gangs or psychotic killers than anything. Steve assured her that this place was safe—a lot safer than Los Angeles, anyway. They went about their business, preparing for the coming darkness, when Steve got a strange feeling, a feeling he had felt fourteen years earlier at Lake Mojave. He turned suddenly in the direction of the nearby mountain, combing its base visually to see who was watching. There was no one. As he lifted his eyes, however, he saw something that had haunted him since his experience back in 1975. There, at Lake Mojave, he and his family had met his best friend's family for a camping trip. He and his buddy, Keith, had gone fishing in the lake at night, only to be terrorized and chased by a large light that came up over the horizon. Steve's only memories ended at the point where he had stopped to help Keith escape, and had been trapped in a light falling on him from the dark sky. He found himself wandering into camp awhile later, terrified—a feeling that never completely went away in these fourteen years.

As soon as Steve saw the light up behind the mountain, it dropped below the ridge, as if to hide from him. This was unnerving to Steve. He attempted to conceal his fear, but Dawn noticed a change in him, exciting her own uneasiness. Steve engaged Dawn in conversations about the beautifully clear, starlit sky, and demonstrated how the desert was teeming with animal life by throwing a hunk of bread into the nearby brush, and watching as a small herd of kangaroo rats pounced on it and carried it away in small pieces. The desert was filled with the chirping and cooing of its inhabitants. Because the stars were so brilliant, Dawn induced Steve to point out the constellations to her. Steve named them one by one as Dawn tested his knowledge. At one point she asked about a particular cluster as she mentioned how it was the brightest in the sky. Steve was astounded as he turned to see the assembly of nine unfamiliar pinpoints of light, near the ridge where he had spotted the ominous light just moments earlier. The group of lights was very bright, radiating light, and was within a few hundred yards of them. Steve was panicking inwardly, as he made fumbling excuses about weather balloons to his questioning wife about why the stars had begun moving in unison across the sky. They watched in amazement, wondering about these brilliant objects, until Dawn announced that they were now blinking to one another as if in communication, and were definitely not stars or balloons. Because of world conditions at the time, the Hesses thought that this may have something to do with nearby military bases. Just as the nine flashing objects floated directly over the Hesses, they disappeared—in an instant they were just gone. Steve and Dawn rapidly searched the starry sky, finding the shining objects now to the west, and several thousand feet up, in the shape of the letter "M."

This was too unusual, even for unknown military maneuvers. Steve and Dawn looked at each other and began to discuss the implications of what had just occurred, when they both looked up at the same time. Hundreds of the flashing lights were now above them, communicating back and forth, blinking out secret messages. Now, the Hesses knew they were seeing something they weren't supposed to see. As they watched in unbelief, they were numbed as the objects began, one by one, to fall gracefully from the sky and land in the desert.

In terror, fearing a full scale Russian invasion or something to the effect, they looked on as more of the orbs floated to the desert. Soon, the desert was filled with thousands of small white lights, moving toward the Hesses' campsite. Incomprehensibly, the sky was now completely black, devoid of any stars, and the desert background noises had ceased entirely. Steve ran to kick out the campfire and ordered Dawn into the back of the pickup. He followed her in, seizing his deer rifle in one hand and shotgun in the other. Then he loaded his rifles, put on his vest with plenty of ammunition, and waited for the fight to start.

As the thousands of lights fast approached, Steve took aim, and was ready to shoot, taunting them with his screams to "come and get it." Dawn, however, was trying to calm him, already receiving telepathic messages that they must surrender or die. With Dawn's guidance, Steve soon heard the "cease and desist" message and felt the will to fight the attackers drain from him. Steve, perplexed by this new pacifism, turned to look out of the back of the canopy at Dawn's prompting. There, he saw two cylindrical forms, approximately three feet high and two feet in diameter, just materializing, but becoming only semisolid. There they oscillated between material and translucent, bluish-grey, making a noise like static electricity, and reeking of pungent sulphur.

Steve lost his balance, falling toward the objects. To his surprise, the nearest of the two floated toward him in like fashion, at the same speed and within the same distance. He sat back quickly and the hovering sentry moved away from him correspondingly. As Steve and Dawn looked out the back of their camping unit, their attention was drawn beyond the two sentries, and they saw that the thousands of lights that had fallen from the sky and raced across the desert had now stationed themselves around the campsite. As they surveyed the sea of glowing objects they knew they were in real trouble—each of them had a pair of red, glowing eyes, starring at the campers with malice.

Then, the small "creatures," translucent with manlike faces and monkeylike bodies, began to frolic around the campsite and surrounding desert, swinging through the mesquite trees and tumbling in the sagebrush. Their play was mocking and malicious—they seemed to be "evil." As Steve and Dawn looked on in numbed awe, Dawn finally broke the silence, voicing her conclusion, "We're going to die."

The Hesses fought to retain control of their minds—it was obvious that the "voices" in their heads were exerting some kind of control over them—but they couldn't tell how much of what they were seeing was real or imagined. Steve told Dawn to pinch him, and they engaged in an ongoing reality test, asking each other questions about what they were seeing, hearing, and smelling. The check established that they were both perceiving the same things.

Steve and Dawn began, the best they could, to analyze what must be happening to them. They considered a military explanation, a psychological interpretation, and anything else they could think of. Nothing made sense. They had begun hearing one another's thoughts, as though tuned in to a central telepathy system. They noted the details of their experience. For example, these little malevolent creatures had to be weightless because as they frolicked through the branches of the mesquite tree, the limbs remained rigid—not even the thin branches bent as the creatures swung on them.

Steve leaned out the back of the truck, shouting at the creatures. As he inclined toward the sentries, one came near him. He reached out his hand and touched it, only to receive a burning electrical shock. The dwarfish creatures appeared to become more excitedly malevolent as the Hesses' terror grew. The gremlins teased and taunted with evil delight as Steve and Dawn held each other in the back of their pickup.

Then, the terror was heightened as they watched a gigantic UFO descend to the earth, hovering about three hundred feet above them and filling the entire valley, blackening the sky beyond the mountain peaks. The craft was the classic disk shape, with a dome in the center on top, and flashing lights blinking in rhythms and codes around the perimeter. They watched in awe and horror as there suddenly appeared six miniature versions of the craft on its underside, and a huge cylindrical light beam shot out from its belly, "beaming" objects into and out of the main craft.

The beam was at least two hundred feet in diameter, and they saw what appeared to be forms in the shapes of living things—not the actual living things themselves—being beamed up and down in the light. The dark silhouettes of deer, cactus, donkeys, mice, trees, bushes, and cattle were floating up and down in the bizarre spectacle. Just then, a lighted form appeared on the underside of the craft. A yellow, red, and white triangle—detached itself and floated to the desert floor, where it began what appeared to be a search of the area, one point down, meticulously combing the basin. The light emanating from the triangle would normally be too bright for human eyes to observe, yet Steve and Dawn looked directly at it, wondering how they could stand the brightness.

They then felt a rumbling within the earth beneath them. They concluded that the triangle must be acting like a drill of some kind, wondering why the triangle might be drilling in the desert so far beneath their campsite. They continued their reality testing, asking each other what they were perceiving. The smell of sulphur had by now become so overwhelming that their breathing was difficult, accentuated by the adrenalin pumping through their veins. As they held each other, contemplating the outlandish scene that had paraded before them for the past two hours, they anguished at what appeared next.

Steve recognized these creatures and knew they were the ones sending the telepathic messages. There were nine greys, the standard spindly bodied type with their large, black, almond shaped eyes wrapping around bulbous heads. Not much in the way of mouths, noses or hair. Only these had an eerie glow to them. They encircled the camper and began their game of mental torture. Dawn suddenly "knew" that these were here to keep her and Steve prisoner, while other greys picked up their children from Grandma's house. Dawn could hear the children screaming for help, pleading for their mother's intervention. Steve could feel the cold observation as the beings scanned his mind, observing him like a laboratory specimen—less than a living being.

Regaining a small degree of his former will to protect himself and family, Steve suddenly burst out with murderous threats against their captors, who sent him reeling with what felt like an "explosion" within his brain. Dawn, panicked, held him, and asked what had happened. He only got out half a sentence when the "experiences" began. Suddenly, full sensory visions, complete with all of the attendant feelings, filled Steve's head. Experiences from his life were relived in rapid succession: the praise and glory of the game winning touchdown; the first time he saw Dawn; hunting with friends; unwrapping Christmas presents—all with the sights, sounds, smells and feelings of the real occasion. Then the pleasantness disappeared, and other sensations and emotions were explored. Steve lost all contact with Dawn and the world, and was completely controlled by the greys, starring into each others' eyes. He was transported to scene after scene, each evoking strong emotions—love, hate, anger, contentment, humiliation, jealousy, pleasure, and pain. The greys drew it all out of him with callous precision, coldly starring as they forced Steve through a lifetime of emotional highs and lows.

Then they made Steve relive the terror at Lake Mojave. They were the ones who had chased him unmercifully that night fourteen years before. They were telling him this—they wanted him to relive his horror and to know that it had been them. Steve began to come out of his trance now, but Dawn had already begun her mental rollercoaster ride. Steve tried to help her fight them, but to no avail—she was already gone, reliving her emotional highs and lows, suffering for their pleasure.

Dawn received the same emotional treatment as Steve, climaxing with the birth of her first baby, Steve Jr. She relived the sheer agony and trauma of childbirth, and then was drawn into the hospital corridor with the sense that something was wrong with the baby, hearing an infant's crying at the other end of the unit. The cast of characters included Steve, their four parents, and some hospital personnel. The drama was executed with perfection by the players—the subtle, suggestive movements, slowly bred fear, then terror, culminating in finding Steve Jr., three years old, in the Operating Room, surrounded by masked surgeons. He called out to her for help, accusing her of abandoning him. Then she saw that something horrible was being done to him. His chest cavity was opened and clamps and IVs were hooked up to him, with blood spurting freely from his arteries. The caustic odor of sulphur was heavy in the sterile room. She then saw that the surgeons were actually the greys, and Steve Jr. cried out, "They've taken me. Used me for experiments."

This scene was accompanied by a deep, menacing sense of loss that took Dawn close to death. Steve and Dawn were being tortured, not only psychologically, but physically. Their body functions, especially their temperature and nervous systems, were fluctuating. The beings were experimenting. Breathing was labored; they were both struggling, trying to stay alive. Just as they both were slipping into unconsciousness, chests heaving trying to take in oxygen, a white fog rolled into the campsite from somewhere in the desert. It came to the back of the pickup and into the camper. It was freezing cold, but as it reached Steve and Dawn, it had an immediate effect on their body functions. It controlled their breathing, like a force on the outside of their chests. It calmed them, instantly, alleviating their high state of anxiety. They were spared, for now at least.

As the wintry haze receded, the Hesses were relieved of their torture and came to their senses, discussing what was happening. They surveyed the activity outside, which had not changed; the triangle still made its unending search for something unknown, and the rumblings beneath them in the earth continued. The greys stared with chilly indifference and the gremlins scurried about in their raucous play. The Hesses discussed the possibilities: were they mining rare ores for their fuel or industrial needs, or did they have another mission? "Something to do with souls," Dawn intuited. She didn't know where the thought came from.

As they reasoned about their captors and their possible motives, Dawn was suddenly pulled again into another game of emotional rollercoaster. This time they knew what was coming and fought to block the greys out, but to no avail. Steve tried to help Dawn concentrate and block them, but the disconnected visions flowed; pride, love, terror, and humiliation kept coming. Then, the most repulsive of all memories—the stranger in the school yard, asking for help, telling young Dawn that her daddy had said it was alright. She followed him into the nearby wooded area where he raped her. Every emotion—the pain, the trauma—it was all there. Only this time, as she came to, she realized that it was Steve's face on the rapist. She recoiled from him, just as the beings began to control his actions. A sneer and leer came over him as he made an inappropriate suggestion that this might be a good time to have sex. Dawn, still reeling from the rape she had just suffered by a man with Steve's face, was astonished as she looked at the greys just standing and observing blankly. Dawn was enraged at Steve's gross insensitivity.

As Steve came to his senses, he found Dawn screaming at him. His apologies went unnoticed. Then, as Dawn was calmed by the vapors of the frosty mist, Steve began to recede into his next psychic whirlwind. He suddenly found himself in a beautiful Pacific Northwest forest early in the morning, wondering what he was doing there and having no recollection or intimation of his torment. He breathed in the nippy, fragrant air, feeling the sun's rays on his skin. His appetite controlled him; he had to find food and answers.

He suddenly smelled a campfire, and followed the odor through the trees and brush. Everything was exactly as one would expect it. He felt very much at home in these surroundings. As he approached, the campfire smoke blended with the scent of cooking rabbit. He was ravenous and had to eat no matter what. He neared the camp but was cautious—there was something abnormal about it. The fire was neatly laid with the roasting rabbit above it. But there was no other sign of anyone having been there—no sleeping bags or gear or even a footstep in the leaves and twigs.

His voracity overcame his sense of propriety, and he took the rabbit and ate. The sensation of satisfaction of appetite was unparalleled. He tore at his feast until he heard a small crackle nearby in the forest. He tossed the remainder of his meal into the brush and headed through the trees, up to the ridge. He suddenly felt watched, as thoughts of hunters baiting their quarry swirled through his brain. Instinctive fear surged through him as the feeling of being hunted overtook him, but it was too late. At the same time that he heard the crack of a rifle he felt the searing rage of a bullet thundering through his left leg.

His entire body went numb as the bullet threw him spinning to the ground. The pain and shock were real, the trauma exacting. He spontaneously jumped up and rambled helplessly through the brush. He was being hunted and had to escape. Questions plagued him as he focused all of his energies on surviving the hunt. The throbbing ache in his leg held him back, while his struggle for his life propelled him forward. He stopped and tore a strip of material from his shirt and applied a quick tourniquet to stop the deluge of blood from his wound.

As he raced off again, dragging his leg, Steve's hopes shattered as he heard the baying of dogs on his trail. His instinct drove him, and he headed for the small river. He splashed out into it and headed straight across to the bank. He made a few tracks and returned to the river, upstream about twenty yards and went for a tree with adequate cover. As he climbed he heard the dogs arrive at the river. They searched for his scent, but had lost it. The hunters soon arrived and cheered the dogs on. The lead dog found his way to Steve's tree as a hunter took aim and shot. The bullet tore into Steve's right shoulder, shattering the bone and liquefying the flesh. He fell from the tree.

The dogs rushed him, tearing his flesh. Steve's mind was fading, thinking thoughts of his wife and children—how much he would miss them. The hunters approached, one of them taking his hunting knife and slitting Steve from sternum to pelvic bone, letting his warm insides roll out over him. Steve faded completely as they began skinning him.

Steve came out of this horrendous vision/experience groggy and frightened. Dawn calmed him as much as she could. Five and a half hours had passed since the stars had begun to fall from the sky.

Just as the two felt they could no longer take this torture, signs of relief came. First, the gremlins began to recede into the desert—first dozens, then scores at a time; then the greys walked out of sight. Steve and Dawn were jubilant there in the back of the pickup, thanking God for answering their many heartfelt prayers for them and their children. They were going to live!

This jubilation was quickly quenched, however, as the greys returned to observe this manipulated ecstasy. The gnomes also returned, this time attacking the camper, creating as much hellish terror as possible. This enraged Steve, and he reached for his shotgun, reviving his sense of safeguarding family. He was filled with fury as he vowed to blast them all. But then, the voice began working on him as his passion was drained by an unseen force. It assured him, "Don't do anything to hurt us. Put the gun down. You have no chance." And then, "In the end, we'll kill you if you try to harm us."

As Steve and Dawn sunk in despair, more greys arrived, watching the back of the pickup. A procession of dozens of them passed by, observing in frigid silence. The great column of light that had been beaming life forms up and down for several hours now neared the truck, and the rumbling in the earth intensified and began to shake the pickup as the beam approached. Steve and Dawn both felt at the same moment that the time had come for them to be taken aboard the great craft hanging above them.

As the beam neared the couple, the truck shook and rocked violently, tossing them about inside the camper, emptying the camping containers onto the Hesses. The stench of sulphur and burning metal permeated their lungs and stung their eyes. The gremlins taunted with delight, threatening and pressing their faces to the windows mockingly. The greys watched, taking it all in. Panicking from their apprehension of what was to follow, Dawn gasped as one of the distant greys lifted its arm, pointing one of its three fingers at her. A small light, the size of a used piece of blackboard chalk, came hurtling through the air, through the window, and into Dawn's abdomen. Dawn tried to reach inside her clothing to brush it off, but it wasn't there—it was inside her.

As Steve and Dawn dealt with this horror, their attention was again diverted to the column of light, which had engulfed them now. The stench, the violent shaking, the howling grinding of the noise beneath them were overwhelming as the two screamed with fear of being lifted up into the mother ship.

They suddenly felt an upward momentum and slight swaying, as though they were being elevated. But as they looked outside, everything seemed normal. The desert was still right there, and the mesquite tree and cactus and brush and rocks were all still there. Yet the sensations of rushing air and of being hurled through the atmosphere were present. Their senses told them that the "drilling" sound must have cut a great chunk of the earth right out beneath them, and that they and their surrroundings were being moved to some other place at high speed. Yet, looking out the windows, they could see the distant mountains and the desert floor leading to them without interruption. They were not moving, despite what their senses told them.

There was an inrushing of chilled air, and the noise, the sulphur and drilling all suddenly ceased. The UFO had disappeared, the greys were all but gone, and the gremlins were retracting into the desert, then beaming up like stars falling upward. A point of light descended in the sky off to their left. As it approached, they could see that it was a person, a beautiful woman in a flowing, luminous white gown. If ever an angelic personage had appeared to mortals, this feminine, seraphic being met their every expectation of what it would be like.

She floated to within about thirty feet of the Hesses, hovering lithely above the ground, standing approximately eight feet tall. A soft, reassuring telepathic voice now permeated them: "It's alright. I'm here now to protect you. Be at peace. It's almost over."

With that, the angelic being ascended slightly, and disappeared in an instant, as Steve and Dawn watched in astonishment.

This visitation had changed everything. The aliens were nearly all gone now. Had they been "chosen" for a special experience or a special purpose? Had they overcome a great test, to prove their worthiness? These questions ran through Steve's and Dawn's minds as they suddenly noticed that the cold fog that had numbed their bodies all night had vanished. They suddenly had an overwhelming desire to use the bathroom. Some plastic containers would have to do—they weren't ready to exit the protective camper just yet.

Sleep was overtaking them hastily as they watched the last of the gremlins beam up to a central point of light that took on the shape of their absent moon.

Steve and Dawn awoke about four hours later at 8:00 a.m., Steve guessed. His watch was not working. Dawn manned the truck's radio to see if there were any reports of large objects in the sky while Steve searched for physical evidence of their visitors. Dawn was disappointed that there was no talk at all about a UFO invasion. Steve scoured the campsite and then the surrounding desert: not a single footprint or broken twig. There was no sign that anything had occurred at all.

Steve started up the pickup and they took off in search of witnesses. They came upon some elderly campers about six miles away, but they had gone to bed early and hadn't seen anything unusual. As the Hesses headed back toward the highway to find a phone to check and see if the children were alright, they watched in astonishment as the triangle, the illuminated "searcher" from the night before, came racing through the sky and followed them. They scrutinized it carefully as it flanked them while they speeded toward the highway. *Will this nightmare ever end?*, they wondered.

A telephone call from a desert cafe assured them that all was well with the children at Grandma's house.

The Hesses returned to their home and children and tried to live a normal life. However, they soon found that the nightmare was not over. In the months that followed, they were plagued with "greys" and other visitors in their home, who took them aboard alien craft and performed horrendous physical and reproductive experiments on them. Poltergeists and other paranormal phenomena afflicted them and their children. They received "revelations" from the beings who had the apparent ability to monitor the Hesses continually. The beings were, Dawn learned, acting under the direction of the "One Supreme," whom the entities told her was God. In these subsequent encounters, the Hesses were terrorized, flattered, told their children would be killed, and told their children would be great world leaders. The mental and physical torture continued, only now it was spread out over time.

The Hesses didn't know what to do about this new phenomenon in their life. They confided in their parents, who received the information with mixed reactions. They confided in a friend, who eventually put them in contact with trauma researchers and regressive hypnotists. Research was conducted and hypnosis sessions held. The Hesses came to believe, as a result of their sessions, that they had not slept for the last four hours in the back of their pickup, but were in fact dragged kicking and screaming aboard the UFO and brutally examined and experimented upon.

Steve eventually joined Dawn's church, and together they sought spiritual guidance, having come to believe that the "beings" may not be extraterrestrial after all. As soon as spiritual protection was invoked, the encounters and paranormal activity ceased instantly. That was approximately a year and a half before this writing, and the abductions and hauntings have not returned.

Eddy Page

Eddy Page was born in 1953 to a woman who had received a complete hysterectomy just seven months previously. Within two months of having all of her reproductive organs removed, Eddy's mother became pregnant with him. Eddy's father, who had been "sensitive" to his wife's condition and had not approached her sexually during that period, was distraught that his wife was pregnant. He sent her away, pregnant and alone. Eddy was born about three months later (at the fifth month of the pregnancy), and looked "inhuman" for the first several months of his premature life. His doctors didn't give him a chance of living—but he did.

Eddy was adopted, which fact he did not learn until he was twenty-nine years old. His adoptive family was kind and supportive, but a schism formed between Eddy and his adoptive father when it was time for Eddy to enroll in college. He decided to enlist in the Marine Corps. During his youth, Eddy experienced anomalous phenomena, but did not realize how unusual his experiences were. Now that he was an under-aged Marine, he felt that he was on his life's path.

Eddy was trained for intelligence work, and was instructed in all of the techniques of self-defense. He was taught well the art of killing humans. He was so good, in fact, that upon his arrival in Vietnam, he was assigned to a special team of twelve assassins who were rushed off on a secret mission to assassinate the President of North Vietnam. His group was transported by plane and submarine and delivered into the hands of South Vietnamese soldiers, who escorted them to the presidential compound. Their objective was to enter the compound, kill the President, and retreat the way they came into the city and country. Their chances of survival were low, however, which they had been frankly told from the beginning.

Eddy befriended a young black Marine during his training and this mission, and they became quite close. When it was time to pair off to assault the presidential compound, Eddy and his friend became one of six two-man teams. Still dressed in their special forces anti-infrared black camouflage fatigues, the six teams invaded the compound. It was Eddy's objective to raid an upper section of one of the buildings in the compound.

All of the teams penetrated the compound successfully, and no shots were fired. Eddy recounts through tears, that he killed his first human—a male military guard, as they moved toward their objective. They made their way upstairs, and as they reached the top, gunfire erupted elsewhere in the compound. People came running out of one of the upstairs rooms into the hallway. Eddy reacted as he had been trained to do, spraying machinegun fire throughout the area, and was sickened as he watched a woman and three children fall lifeless to the floor. His friend heard commotion from a room they were near and tossed in a grenade. The room exploded and a ball of flame came roaring through the door.

Eddy and his buddy suddenly had a very strange feeling come over them, and seeing that no one else was present in their targeted area, they retreated back down the stairs. By this time, a firefight was ensuing in the compound, and just beyond the compound they could hear what sounded like an entire army coming in on them. Eddy and his friend turned a corner and encountered a spray of bullets. His friend went down, bleeding profusely from a wound in his side. Eddy was stunned, and stupidly stood there and looked into his buddy's eyes as he immediately sank into death.

Eddy sprayed the foreground with bullets, but was a standing target. He looked back down toward his dead friend lying at his feet, but to Eddy's astonishment, his friend was not there. He looked quickly around and found that he was no longer in the presidential compound—he was standing in the middle of a rice field. Three American helicopters were just breaking the treeline in front of him. He glanced back down to see where his buddy was, but found only water and plants. He also noticed that he was no longer wearing his special black camouflage fatigues, but was dressed in fresh, pressed green fatigues.

As men jumped from the helicopters to retrieve Eddy, his mind whirled with questions: where was he? where was his buddy? where was the compound? He didn't understand what was happening to him.

Eddy was flown to a nearby outpost where he received a cursory debriefing, and some very unkind treatment. He learned that some time had passed—days or weeks, he wasn't sure—for which he could not account. He was soon flown out to a military base, where he was checked out physically, and put through horrendous interrogation. Accusations of conspiring with the enemy were leveled against him with venom. Threats of prison and execution were hurled at him. He was treated as a traitor, and feared his American captors.

Eddy was approached by a doctor who was running tests on him, who told him that he was sending Eddy to the States. Eddy thought this meant to his family and friends, but the doctor informed him it meant Bethesda, for significant medical examinations. The doctor was comforting and told Eddy not to worry—he would be alright. He told Eddy that he was special.

So far, not a single question had been answered. Eddy wondered why he had gone into battle with his friends and for his country, and was being treated like an enemy. He wondered why he was standing in gunfire one second and a rice field the next, dressed differently. How had the American choppers known where he was and when he would be there? How did they obtain free passage into and out of enemy territory? Or, was it enemy territory? In fact, Eddy couldn't even be sure of where he was when the helicopters had arrived. Eddy was certainly not receiving any responses to the questions he put to his superiors.

Again, at Bethesda he was grilled by an accusing, overbearing intelligence officer, making every effort to emotionally batter Eddy into a confession of how he had betrayed his unit and country. At the same time, lasting for months, Eddy was the subject of a great deal of medical testing. Fluid specimens were collected around the clock for analysis. EKGs, X-Rays, and other tests were administered. For months this went on. Eddy had no idea what was occurring, but he knew that he was the center of something colossal.

As before, it was the doctors who comforted Eddy and let him know that he was in no trouble. Again, the word *special* was used. Eddy was special, but he had no idea why. Finally, the brutal intelligence officer came to Eddy and offered him the opportunity to transfer to a new assignment in the Corps or to receive an honorable discharge and go his own way. Eddy had entered the Corps with the intent of making a career of it, so he opted to take a new assignment. He stayed in another six or seven years, receiving accelerated rank advancement because of his outstanding performance. At that time, however, he was told that the Marine Corps no longer had need of his specialty, and he was asked to resign. This was quite puzzling to Eddy, because his specialty was motorpool mechanic. Unless the Corps had planned to immediately phase out its use of motorized vehicles (which did not occur, of course), Eddy's specialty was needed.

The message was clear, and Eddy was honorably discharged. He received at that time (not before), his many medals and citations for his heroic service in Vietnam and outstanding service in the Corps. Eddy left the military, still wondering what it was all about, not having received a single word of explanation for what had happened in Vietnam, or what they had discovered at Bethesda.

Eddy took a job in Florida, and within a couple of years married. His young wife became pregnant, and one evening when she was nine months along, something peculiar happened to them. Eddy was driving down the freeway, and without explanation suddenly stopped the car and got out. His wife called to him to find out what he was doing, and was shocked to see a reddish-orange "craft" appear in the middle of the freeway. She looked on as Eddy approached the craft, and was startled that in the next instant he was opening the car door and getting in.

She asked Eddy what he was doing. Eddy was confused, and wondered why he was not still driving down the freeway—his last recollection of what he had been doing. He said the first thing that came into his mind—that he had to urinate, and had stopped for that purpose. His anxious bride was coming unglued and demanded to know what that craft was that Eddy had walked toward, and how he suddenly appeared at the car. Eddy had no answers for her. She went into labor in the next few hours and Eddy took her to the hospital. She insisted that he be kept from the delivery room and from the baby. Eddy never saw his baby girl or his wife after that. She divorced him and never allowed him to come around again.

This was as confusing as any part of Eddy's life, and his loss of wife and baby daughter grieved Eddy tremendously. He searched for answers in the months that followed, and learned then that he had been adopted. He looked up his biological parents and was received with mixed emotions.

He soon experienced physical pain as well when his brakes went out at an intersection, leaving Eddy pinned in his car under a gasoline tanker. He was extricated and rushed to the emergency room of the largest hospital in the area. There, lifesaving procedures were administered and longterm damage assessed. He was told by his doctor that his cracked forehead would cause him a great deal of trouble in the weeks and months that followed, and that multiple surgeries would be necessary before he could walk again on his shattered knee.

Within days, doctors were again buzzing around Eddy's hospital room, performing examinations and running tests. His treating physician came to him and asked how he felt. He responded that he felt just fine—great, in fact. The doctor then proceeded to recount how they thought they had mixed up his x-rays with another patient's file because the latest set had shown no sign of the fractures captured on film just a few days before. This was impossible, but they had now confirmed that the fractures and the broken ribs were completely healed. The doctor asked Eddy about his previous experience of being shot. Eddy stared blankly, asking what the doctor meant. The doctor said Eddy had been shot previously, and showed him an x-ray disclosing a bullet still lodged at the base of his skull, imbedded near the brain stem. Eddy insisted that he had never been shot.

The doctor then asked Eddy if he could examine his as yet untreated knee. The doctor exposed the leg and removed the splint. He gently checked for any degree of mobility in the knee. It moved full range with only minor stiffness. He was amazed, and exclaimed that this just wasn't possible. Full tests were ordered, and the conclusion Eddy's doctor proffered as he walked with Eddy to the hospital exit two days later was that Eddy was not a normal human—in fact, there was little indicating that Eddy was human at all, except that he looked perfectly normal.

Eddy was discharged in perfect physical condition about a week following the devastating auto accident. Again, he left with more questions remaining than answered.

Within several months, Eddy met another girl and remarried. His new bride was pregnant within a short time. They lived in Alabama because Eddy had found a well-paying job there. All was not well in paradise, however, and Eddy's wife complained that she was frightened of Eddy. Other personages were talking to her through Eddy, telling her things about herself and her new little family and telling her things about their future. She was quite frightened at this, yet Eddy had no recollection of these "channeling" sessions.

She finally left Eddy and filed for divorce. Eddy's son was born and again, he did not get to see his child.

Eddy sought psychiatric help at the Veteran's Hospital, revealing what little he knew about himself, and receiving all of the standard psychological testing and profiling. His doctor discharged him within a few days, telling him that he was mentally healthier than many members of the staff. Eddy learned what he could about himself through his adoptive and biological families. At this point, he learned about his mother's hysterectomy and subsequent pregnancy and premature delivery. Putting these facts together with the craft that his first wife had seen him approach, and the periods of missing time, Eddy sought counselling from UFO investigators. They took Eddy through regressive hypnosis wherein much information was obtained.

The most pressing questions were those related to the Vietnam experience. This was a gaping hole and burning question in Eddy's life, and he had to know what had happened to him. In his videotaped hypnosis sessions, Eddy relived the brutality of that day in the presidential compound—shooting the woman and children, and looking into his buddy's dying eyes with gunfire hailing down and spraying all around them. Eddy was hit, taking at least seven bullets. His uniform was riddled with bullet holes and Eddy was entirely disemboweled. He went down, having only seconds of consciousness remaining in his obliterated body.

He suddenly was enveloped in a bright light, and found himself in the presence of the beings—the greys. They took him—where and how he wasn't sure—he only knew they took him. The pain was excruciating, but he was fast fading. The greys lowered a metal-looking cover over his eviscerated body and the pain stopped immediately. He was transported to another area, accompanied by greys, one holding his hand and giving comfort the entire time.

Eddy was then placed into an upright, transparent cylinder filled with a liquid of some kind. He found that he could breathe the liquid and just floated there for several hours, feeling no pain. When he was first placed in this tube, however, an interesting thing happened. The grey that had comforted him earlier walked up, and signaled encouragement and affection, smiling at Eddy, to the degree greys can smile with the small slit of a mouth that they possess. He then reached forward placing his two extended fingers on the surface of the tube. Eddy intuitively responded, and as their fingers met on the inside and outside of the thin transparent wall, Eddy was filled with an overwhelming sense of well being, comfort, and even ecstasy.

After several hours had passed, Eddy was removed from the cylinder and escorted, healed and naked, through the craft, or whatever it was he was in. He was amazed that he was completely restored, with no sign of a wound anywhere. The comforting grey accompanied him, showing him around the craft. They came to a very long corridor, at which point the grey introduced Eddy to a female grey, who took over the tour. She led Eddy through large doors into large rooms, where Eddy saw many greys, going about their tasks. Eddy was naked and was becoming cold. They had apparently seen everything Eddy was allowed to see, so the female returned him to the corridor where he waited alone for awhile until the comforting grey returned.

The two spoke, telepathically, and the grey informed Eddy that he was disappointed in Eddy's activities on the earth. Eddy thought his life was following a noble course and asked what the being meant. He responded that he was deeply grieved that "a son of his" would ever kill another creature. With this revelation, Eddy was made to believe that this grey was his biological father. The grey then showed Eddy that the very long corridor that they were in was lined on either side with small transparent containers, in which little "babies" were kept. Eddy remarked that they looked like descriptions of him when he was a premature baby. The grey then pulled a picture out of his breast pocket, and showed it to Eddy. Eddy looked at the picture, which looked like one of the babies in the containers, and was told that it was a picture of him.

Eddy and the grey said warm goodbyes, and Eddy next found himself in the rice field with American helicopters coming over the treeline.

After contacting the UFO investigators Eddy's outlook changed, and his second wife contacted him, expressing an impulse to be with him. The UFO investigators are still working with Eddy, and are in constant contact with his father, the grey, through Eddy, who channels him. Eddy feels rather left out of this process, because he is not conscious during the sessions. Recently, Eddy became angry about this situation and demanded to receive a communication from his space father.

The next day it happened, in the back yard. Eddy's neighbors had slowly learned that he and his son now periodically see UFOs. In fact, Eddy is informed when the UFOs are about to make an appearance, and is usually told to bring a camera. In fact, if he does not obey these promptings, he is chastised—told to go buy a camera, or not to be late. On occasion, Eddy invited his neighbors to come out and see the UFOs when they come. On this morning, those neighbors just next door to him were watching out the window as Eddy had a visit from his space father. They only observed the dog walk over to the fence, and return to Eddy looking up playfully as though he could see someone walking with him. All they could see was a light that came and enveloped Eddy.

Eddy's encounter with his father lasted about half an hour, and was very satisfying to Eddy. At one point Eddy's father reached out and touched Eddy on the shoulder, and Eddy did likewise, feeling his body and skin.

Eddy's body is different in many respects than most humans, according to him and some doctors, purportedly. Apparently, Eddy's internal organs are physically different and his functions are unlike other humans. He was recently able to "steal" a copy of part of his service record. The single page he obtained was his internal discharge form. UFO insiders are buzzing about this document. One interesting aspect is the reason Eddy was forced to resign. The document says:

> TO STAY ON ACTIVE FILE UNTIL DISCLOSED BY 99 WITH NAVAL INTELL....74....PRESENT BEING OF UNHUMAN PRIORITY TO BE WATCHED BY ORIPO....FOR MJ-12 ONLY.

Eddy was a being of "Unhuman" priority, according to the discharge form. The exciting part of the document, however, is the instruction to keep track of Eddy, have him monitored, and to report to "MJ-12." MJ-12, or Majestic-12 as it is also called, is a "highest level" government (or multi-government) "black" organization that actively investigates and covers up alien encounters, according to those of the conspiratorial belief. Many believe that MJ-12 not only knows all about UFOs and their extra-terrestrial pilots, but is actually participating at some level with the aliens. The discharge document is important because it is one of the few possessed by civilians pointing directly to the existence of MJ-12, according to UFOlogists.

8
Contact

Early Contactees

In the early decades of the UFO "invasion," there were those who came to the fore with tales of extraterrestrial contact, bearing messages of intergalactic peace and Earthly holocaust if we failed to heed the UFO message. These "contactees" were generally thought to be crackpots. That perception still persists. Their messages were so unbelievable, and their demeanor so incredible, that the public found no reason to heed their message. However, as incredible as it may seem, it could well be that these contactees were accurately reporting their perceived experiences. The most important question is, who contacts the contactees?

George Adamski

The first of these galactic messengers was a Polish immigrant, George Adamski (1891-1965), who called himself Professor G. Adamski. He listed his address as Mount Palomar, California, where anyone would assume he was associated with the famous Hale Observatory, home of a 200-inch telescope, the largest in the world in 1944. Although it is true that Adamski lived on the southern slopes of Mount Palomar, seven miles from the Hale Observatory, his closest contact with the high tech folk was to sell them coffee from his roadside stand.

Adamski wrote science fiction space books, which undertaking was insufficient to pay his bills—hence, the coffee stand. In *Flying Saucers Have Landed* (1953), Adamski claimed to have seen his first UFO in October, 1946. He claimed he saw 184 UFOs in a single night in August of 1947, coming and going at will, streaking busily through the California night skies. Adamski wrote of flying saucer rides to the planets in our solar system, including the dark side of the moon with its rolling green hills and small villages hidden from our view. He further claimed to have encountered many Venusians who looked like humans. Interestingly, in the early 1950s it was still widely believed that the other planets of our solar system were capable of sustaining life and were the source of extraterrestrial visitation. Later, following scientific scrutiny of the surfaces and atmospheres of these nearest planets, contactees began to report that the aliens were from the nearest star systems instead.

Adamski's Venusian friends claimed to be quite concerned about radiation leaking from Earth's atmosphere. The alien elder philosopher, The Master, discussed many galactic concerns with Adamski, the major consideration being Earth's perilous threat of destroying itself and nearby planets by its tinkering with nuclear power. The Venusians further explained that they lived among humans, and monitored us thoroughly. Aliens from most of the planets in our solar system had representatives on Earth, and because they look like us and live among us, he encountered them frequently in cafes and bars throughout greater Los Angeles.

The Aliens explained that Jesus Christ had been their spokesman at one time—now Adamski was selected to bear the Universal message of our "space brothers." This Galactic Messiah sold many books and worked the lecture circuit heavily, giving birth to an American industry. A flurry of contactees began lecturing and writing on their similar experiences of joy riding through our solar system and receiving special messages and missions from the space brethren.[1]

Orfeo Angelucci

Another contactee, Orfeo Angelucci, published his *Secret of the Saucers* in 1955. The message was wholly spiritual by now, containing prophecies complete with deadlines. His space brother mentor, or contact, gave him interplanetary revelation. Angelucci rode in a flying saucer, met Neptune, and received a mystical revelation, filling him with mystical knowledge. He met with Jesus once, who explained that the extraterrestrials are here to help us, announcing, that "this is the beginning of the New Age." (Remember, this was over thirty years ago.) Aliens explained to him that in a former life, he had been one of the space brethren from another planet. He prophesied of a major world catastrophe by 1986.[2]

Other Contactees

Another 1950's contactee, Howard Menger, saw his first alien, a beautiful blonde woman, in 1932, and was repeatedly contacted throughout his life.

---

1 *Phenomenon: Forty Years of Flying Saucers,* Avon Books (1988) pp. 121-124.
2 *Ibid.*, p. 126.

It was easy to spot the men he explained—they had long blonde hair too. He always described an overwhelming aura of love and harmony when in the presence of the aliens. These space brothers told him that many early Earth civilizations had been contacted by them and had received superior knowledge and technology from them, most of which was lost in ancient wars.[3]

In 1954, George King, a London taxi driver with a background in the occult, heard a voice while washing dishes. The voice told him, "Prepare yourself, you are to become the voice of Interplanetary Parliament." King founded the Aetherius Society, and slipped into public trances during which he acted as a medium for interplanetary communications. King revealed how he had traveled to many planets, fought a megabattle in space, and conversed with many beings on different planets—including Master Jesus, who lives on Venus.[4]

Many UFO gurus came forward in the 1970s and 1980s, most of them eventually moving to southern California with their cultish groupies. Their overt messages were generally that love and brotherhood must replace materialism, war, and nuclear power. They preached of a coming age when man would be elevated to a higher spiritual plane and enlightenment. It would be easy to dismiss these pre-New Age contactees as mere charlatans who fabricated science fiction tales to gain a following. However, certain facts jump out of their stories that compel us to at least notice particular consistencies with both New Age doctrine and more credible UFO contact accounts. In particular, Howard Menger of the early 1950s, before the UFO literature was well-developed, spoke of his beautiful superrace of blonde men and women, describing an overwhelming feeling of love and harmony while in their presence. This is a phenomenon related by many others, including, Udo Wartena. Interestingly, Howard Menger was probably the "tamest" of the early contactees, speaking little of New Age teachings of the space brothers. The remaining contactees were filled with messages closely akin to New Age doctrine. Although such beliefs were held, or at least professed by some in the "fringe" of the time, it is fascinating that most of the contactees recounted their tales in nearly mystic terms.

---

3 *Ibid.*, pp. 126-28.

4 *The UFO Phenomenon,* pp. 78-79.

Assuming that most of these contactees were not really in communication with extraterrestrials (a safe assumption in the author's opinion), we are almost compelled to accept that some of them were in contact with intelligences that fed them false, unreliable or weird information, for reasons not fully apparent at this point in our discussion. It is a safe assumption, the author believes, that their contacts were of the same nature as are those experienced by New Age channelers.

Public Rejection Of Early Contactees

The public backlash was severe against the early contactees, generating general disdain for any person claiming to have seen an alien—or anyone claiming more than one UFO encounter of any kind. Although the public could accept that UFO sightings were truly occurring, the claim that UFOs contain alien pilots was insurmountable. Although a few self-proclaimed contactee/messengers are still to be found, preaching that they are the chosen vessels of intergalactic bulletins, most current contact with extraterrestrial entities is accomplished through New Age channelers—most of whom have at least one alien in their repertoire. The following messages from extraterrestrials to New Age channelers are representative of those published by the hundreds in New Age literature:

- I *am* one of these Advanced Spiritual Beings. I have come from a very high spiritual world called Aries; I am not an Earth Person. I am now living as an Earth Person in a physical body. I came to help the Earth people. [Uriel, through Ruth E. Norman of the Unarius Foundation]
- Will you agree to be the savior of the world? [Ashtar, to American contactee Allen-Michael Noonan]
- I, Raymere, transmit once more upon this occasion in order to speak with you about the things of the next period of time . . . you will find that you are moving into a higher frequency wherein there is a totally new dimension. [Raymere, a space being, through Alenti Francesca at the Solar Light Retreat]
- Earth's vortex is about to break because of an excess amount of hatred . . . . [The space brothers of Io, a moon of Jupiter, to "Gordon," a U.S. contactee, 1967][5]

---

5 *Phenomenon: Forty Years of Flying Saucers,* pp. 365-66.

Contactees—The Next Generation

Although the first wave of galactic gurus has come and gone, leaving in their wake a feeling that rainmakers have beat the drums, fired the cannons, and sold a little snake oil, a new breed of messengers has appeared on the horizon. They are the abductees discussed herein who have apparently not chosen to be contacted, who do not appear to seek their contactee status, but who have been given a message for the world just the same.

The Andreasson Affair

According to Raymond Fowler, longtime UFO investigator and author, the actual events that make up a UFO abduction constitute the components of a composite UFO message. As we examine such events in greater detail, this theory becomes quite feasible. In his book, *The Watchers,* subtitled, "The Secret Design Behind UFO Abductions," Fowler documents the case of Betty Andreasson Luca,[6] who for decades was reportedly abducted by UFOs. Her case includes many of the experiences reported by others as discussed herein, with the added dimension of being given a time-capsulized, piecemeal message to deliver to mankind. The message given to Betty Luca is being simultaneously implanted and released in the minds of others around the world. We examine Betty's experiences more closely in this chapter, however, because of the progress investigators have made on her case and the publicity given to it. The message, as gleaned from Betty's excavated subconscious, parallels the same message retrieved from the memories of many others. It fits certain patterns that warrant our attention and provides important information in our search for the source of UFOs and their message.

Betty Luca is described as a woman having "a deep and exceptionally beautiful Christian faith." The emphasis is on the "Christian" element because Betty's experiences were interpreted by her as "religious" in na-

---

6 Fowler has written three books chronicling this case. The first was phase one, *The Andreasson Affair,* and the second, *The Andreasson Affair: Phase Two.* Subsequent to each of the first two books the subject, Betty Andreasson (Luca), experienced surges of memory recall as "time-capsule" messages that were unlocked in her programmed memory, and as she was further abducted. Fowler assumes he has written the last book on this case, believing that all has finally been revealed. The author doubts this.

ture—at least initially. An added dimension to the Christian aspect of Betty's experiences is that she often sought divine protection at the beginning of many of her experiences, to no avail.

Betty's first encounter with the small greys was at the age of seven, in 1944. She recounts, under hypnosis, in the persona of a seven-year-old girl what happened to her on that occasion.

> I'm sitting there eating some crackers looking at the blue flowers outside the hut, and I'm waitin' for Didi to come over and play. And then all of a sudden I see a bumblebee or something, but it's bright light and it keeps on circling my head. Maybe it's after my crackers. But it keeps on going round my head and then it stuck there . . . . It was cold and it was making me fall backwards and I felt very sleepy. I'm lying on the ground there and I hear something. There is a squiggly feeling in my head, and there is a voice speaking to me. There is a lot of them, but all talking together. And they are saying something. They have been watching me, and uh, I'm coming along fine. And they're talking to me and telling me that I'm making good progress . . . and they were getting things ready. But it won't be for awhile . . . about five years or so. I would be twelve. They would see me later.[7]
>
> [Another session covering this same experience offers a little more detail:] It's coming to a time that I will know the *One* . . . . They're going to show me something . . . that everybody will be happy about . . . that everybody will learn something from . . . . They just want to look me over *from the inside.* They tell me I'm going to be very happy soon . . . that I'm going to find the *One.* I will feel the *One.*

The phenomenon of many voices speaking in unison is common in telepathic UFO communication. Betty learns from the chorus that they have observed her, that they are making preparations for some great event, and that she would be ready for the event in about five years, at the age of twelve. An important component of the event is to go somewhere to *know, find,* and *feel* the *One.* Another component is to see something that will make *everyone* happy, an event instructive to everyone. The parallels to New Age beliefs as discussed later in the book, are obvious if these promises are taken at face value. As we shall see, the realization of these experiences is everything promised, and more.

Betty experienced a second encounter, as promised, when she was twelve, in 1949. She recounts that while playing in woods, she encountered a small grey in a high-tech-looking uniform. Having no conscious recollection of her previous encounter, and not knowing what the ugly little

---

7 *The Watchers,* pp. 7-8.

fellow was, she did the appropriate thing: "I took some of those stones out of my pocket. I thought it was an animal coming out. I started to throw stones at it, and, ah!—The stones hit something and *stopped in midair* and just fell down!" Betty heard the same voices say that she would not be ready for another year.[8] "They said I will learn about the *One*. They said they are preparing things for me to see."[9]

On this occasion Betty appears to have been one year premature when being checked by the beings. Premature for what? By the time of Betty's next encounter a year later, Betty had become a young woman, approaching sexual maturity. Apparently, this had something to do with an event for which things were being prepared. She would now be ready to meet the *One*.

At the age of thirteen, now 1950, following an impulse, Betty got up early one morning while her family slept and went to explore near the pond. She saw a "huge moon" coming over the hill toward her. It became "bigger and bigger" as it came toward her—she tried to run, but was paralyzed. She next found herself inside a white room feeling "very relaxed," watching as two small greys floated toward her a few inches off the floor. They said, "We're going to take you *home*." Betty responded, "I am home!" And they said, "Don't fear, don't be afraid, you're alright."

The One

Betty was placed onto a soft "cushion-like mat on the floor of a section of the craft that was roofed by a large transparent dome. A mouthpiece was installed that kept her tongue held down." The craft accelerated at high velocity and after a time entered water and descended to an underground complex. In the complex Betty went through a museum of time with glass cases containing human replicas in the garb and natural habitat of various historical periods of the Earth. She next underwent a physical examination. She was then told, "You're getting closer to *home*," and was taken to a clam-shell-looking device with mirrors inside, which she was instructed to get into. It closed, and opened a moment later. She found herself in a

---

8 *The Watchers,* pp. 8-9.
9 *Ibid.*, p. 331.

different place made of a glass-like substance. She was shown glass-like replicas of animals and plants that were quite unusual:

> And I'm reaching out to touch a butterfly and when I did, it's fantastic! It's beautiful! There's all color coming into the butterfly now, and it's flying around and around. When I touched it, it got color and lived and it's flying . . . . Oh, it's stopped. Its color is going and it's fading into a tiny speck of light, like a tiny speck of light. Then it goes back into the ice-form of that butterfly. That was amazing! I asked him "What's happening?" He says, "This is for you to remember so mankind will understand." And I said, "But why did it turn color and fly away when I touched it?" He told me that I will see when I get *home.* He said, "*Home* is where the *One* is."[10]

Again, Betty asked how they could do these things, and relates the response: "He told me that I will see when I get *home.* It is for me, they said, for me to go *home* to see the *One.* He said, '*Home* is where the *One* is.' He says, 'We are drawing closer to *home* where the *One* is'"[11]

All these references to the *One* and going *home* were obvious attempts to make Betty understand that she had a close kinship to whatever awaited her—a close kinship to the place and the *One.* After being "transported" to this place called *home,* Betty was taken to a Great Door:

> We're coming up to this wall of glass and a big, big, big, big, big, door. It's made of glass. [Q. Does it have hinges?] No. It is so big and there is—I can't explain it. It is door after door after door after door. He is stopping there and telling me to stop. I'm just stopping there. He says: "Now you shall enter the door to see the *One.*" And I'm standing there and I'm coming out of myself! There's two of me! There's two of me there! . . . It's like a twin. But it's still, like those people I saw in those, those ice cubes [glass cases in the museum].[12]

This phenomenon being experienced by Betty is known as an out-of-body experience ("OBE"). It is the ultimate spiritual experience in the occult/New Age movement. What is interesting is that Betty evidently has done nothing of her own choosing to initiate her OBE.

Upon entering the Great Door during this OBE, Betty recounted under hypnosis what she was *able* to tell. Raymond Fowler reports that during this portion of the hypnotic session "a rapturous, beatific expression of pure, unrestricted happiness came over her face as she apparently met

---

10 *Ibid.,* p. 333.
11 *Ibid.,* p. 146.
12 *Ibid.,* p. 11.

. . . the *One.*" Betty would not or could not describe what happened next. She attempts:

> It's—words cannot explain it. It's wonderful. It's for everybody. I just can't tell you this. [Q. You can't? Okay, why can't you?] For one thing, it's too overwhelming and it is . . . it is undescribable. I just can't tell you. Besides it's just impossible for me to tell you. [Q. Were you told not to share it with me?] It is like even if I was able to speak it, I wouldn't be able to speak it. I can't. I'm sorry . . . .
>
> And I'm standing there and I'm coming out of myself! There's two of me there . and the little person is saying: "Now you shall enter the Great Door and see the glory of the *One.* I went in the door and it's very bright. I can't take you any further. [Q. Why?] Because . . . I can't take you past this door. [Q. Why are you so happy?] It's just, uh, I just can't tell you about it . . . . It's—Words cannot explain it. It's wonderful. It's for *everybody.* I just can't explain this. I understand that *everything is one.* Everything fits together. It's beautiful![13]

Whatever Betty was experiencing, it was beyond her ability to describe to investigators. The experience is very much like the mystical experiences professed by gurus and other spiritualists. It is also very much like near-death experiences (NDE) that we read about in ever-increasing volumes.

The "glory" of the *One* will one day be experienced by all, Betty perceives. She gains the ultimate spiritual insight that "everything is one," and that "everything fits together." Although this concept can be found in various religious traditions, it is also very much a part of the over-emphasized "all is one and one is all" tenet of the New Age/occult religion.

Many contactees are being told that a supreme being, the ultimate "alien" leader, is directing the alien contact with Earth. In Dawn Hess's hypnosis session she was asked questions about the aliens' contact. "Are we watched as a population regularly?," she was asked, to which she responded, "Yes." "By many different kinds of beings . . . ," she finished his sentence, ". . . sent on missions from the One Supreme." She explained, "There's one supreme Being that controls all of them. He sends missions here. They're not here of their own accord." He queried, "This is like a president, or . . . ," again she interrupted, firmly, "Like a god." After some discussion, Dawn was asked, "Do we have a separate god from the *One Supreme?*," to which she quipped, "No." Dawn Hess is a woman with very firm Christian beliefs. During her hypnosis session she was speaking and responding in an uncharacteristically aggressive way, giving those present an elevated

---

13 *Ibid.*, p. 144.

sense of anxiety. In subsequent discussions with Mrs. Hess the author found that she was surprised by the information she revealed during her session, and has no personal conviction that the One Supreme is supreme in any way, or that he is God.

Dr. Mack similarly documents the alien claim that they are on a mission from God. "The aliens are recognized as intermediaries or intermediate entities between the fully embodied state of human beings and the primal source of creation or God (in the sense of a cosmic consciousness, rather than a personified being). In this regard abductees sometimes liken the alien beings to angels, or other 'light beings' (including the 'greys')." (*Abduction,* p. 48.) Jerry, a female abductee with whom Dr. Mack worked related, "She asked [her 2 year old son] how he got onto the spaceship and he circled his eyes with his hands and said, 'The eyes.' When Jerry asked him if there was anyone on the spaceship with him he answered, 'Yeah, I see the King. I see the King and He is God.' Jerry wondered where he obtained the verbal skills, seemingly beyond his years, to say these things." (*Ibid.*, p. 116.)

The hypnotist made several subsequent attempts to get Betty behind the door, but she could not get past it. The "brightness" experienced during the hypnotic attempts gave her eye aches and headaches for several days following such undertakings. In a phase one session, Betty had mentioned the Great Door, but the investigators failed to follow up on it. When asked, "What is the *Great Door?,*" she responded, "It is the entrance into the *other world.* The world where light is." When asked, "Is it available to us as well as to you?" Betty responded, "No, not yet."[14] Although these very unique and intriguing experiences had occurred fifteen years before the investigation had begun, the investigators were unable to get to these "message" portions of the experiences until "the time was right." The messages were implanted in Betty, and others evidently, within timesensitive memory barriers. Even when the time for release of the messages arrived, investigators still encountered difficulties getting to them. They had to invent the curtains, television screens, and other mental protective devices discussed later in this chapter to get into the blocked memories.

After leaving the *One* Betty returned through a tunnel and went back through the Great Door.

---

14 *The Watchers,* p. 145.

> Okay, I'm outside the door and there's a tall person [human] there. He's got white hair and he's got a white nightgown on and he's motioning me to come there with him. His nightgown is, is *glowing* and his hair is white and he's got bluish eyes. And it's bright out here, and I think I see two more of them over there. [Q. Do they look like people?] Um, but tall. They are real tall and they got some ferns or something in their hands. [Q. Do you speak to them?] He's beckoning me to come over there and there's like a *shell,* an open *shell.* But it's mirrors and mirrors and mirrors.[15]

Betty was introduced through the Great Door by small greys, but was received, upon exiting the Door, by what she considered to be angelic humans following her encounter with the *One.* In Betty's accounts, as well as those of hundreds of others, these angelic humans are often seen in close proximity to small greys, and appear to direct the greys. Raymond Fowler suggests that Betty was welcomed back through the Great Door after having been with the *One* by these higher level beings as a sign of honor to the chosen one. Although a chosen one, it did not take long for Betty to be treated like a laboratory specimen again.

Betty was transported through the clamshell device as before, and received at the other end by the small greys. At that point Betty was taken to another examination room where she was floated up on a square table. To her horror Betty sobbed in pain as she recounted how the beings removed one of her eyes and implanted a small round object behind it. They then placed a long needle up her nose and implanted a BB-like object with tiny spines protruding from it up her nasal cavity, in the classic membrane penetrating fashion. Although the nose implant is common in abduction literature, the object behind the eye is not. Betty described a sophisticated procedure as the round unit was implanted and tested. The implication is that all of Betty's sensory organs were wired for continuous monitoring by the beings.

Follow the Son

In 1961 Betty had a "religious" encounter with the beings. She was a 24-year old mother at the time. While mopping and singing hymns, she heard a strange noise. Despite having children napping alone in the house, she walked outside and into the woods without volition. She struggled up

---

15 *Ibid.,* p. 150.

a nearby hill, slipping on pine needles and climbing over rocks. The new message she received during this encounter warrants our attention.

> There's a strange being standing over there and I'm afraid of it. It's staring at me, and I can't move! Oh, Jesus be with me! He's telling me [telepathically] that I have been *watched* since my beginning. I shall grow naturally, and my faith in the Light will bring many others to the Light and Salvation because many will understand and see . . . .
>
> He has been sent and I am not to fear. The Lord is with me and not to be afraid. They are pleased because I have accepted [Christianity] on my own. I am to go through many things and that love will show me the answers because I have given my heart over to love the Son. Many things shall be revealed to me. Things that I [eyes?] have not seen . . . ears have not heard . . . . I shall suffer many things . . . but will overcome them through the Son. I have been watched since my beginning. I shall grow naturally and my faith in the Light will bring many others to the Light and Salvation because many will understand and see. The *negative voices* don't like it. [They] are against man . . . bad angels that wanted to devour man . . . hurt man . . . destroy man . . . because they are jealous . . . of the love that is upon man. Telling me strange things . . . I don't know what they're about . . . That for every place there is an existence. That every thing has been formed to unite. [He says] Jesus is with me . . . that I will understand as time goes by . . . for me not to be anxious. They want me to grow and live naturally . . . that I am blessed and that I will forget and I am now to go back to my house and I will not remember. He says, "Peace be with you as it is."[16]

It is interesting that as Betty perceived the danger of her encounter, even though she was mentally tranquillized by the small greys, her first thought or prayer was, "Oh, Jesus be with me!" At this point, the message became highly religious in nature, almost like a visitation from a heavenly being. The message was compound: the beings have "watched" Betty from infancy; they are pleased she has accepted Christianity; her faith in the Light will bring many others to the Light and Salvation; there exist evil alien forces, "bad angels" who are jealous of mankind and seek our destruction; and, although Betty will undergo severe hardship, her faith in Jesus will pull her through.

According to the account, in January, 1967, Betty was again abducted by the beings and taken to a distant planet. As she travelled between her two abductors along a high trestle she entered a beautiful crystalline structure to witness a Phoenix legend enactment. There was much light and heat, and Betty cried out for help and writhed in agony. Soon the temperature dropped and Betty was able to squint her eyes open. The bird was gone,

16 *The Watchers,* pp. 334-35.

and a small remaining fire dwindled into embers and ashes, from which emerged a "big fat worm." After this Betty heard a "thundering chorus of voices blended together as one mighty voice" calling her by name and asking, "You have seen and you have heard. Do you understand?" Betty responded, "No, I don't understand what this is all about, why I'm even here." The voice instructed, "I have chosen you." Betty again asked, "For what have you chosen me?" The voice answered, "I have chosen you to show the world."[17] Betty felt that the voice was that of God Himself. Although she did not see him, she was sure it was He. Betty professed her belief in Jesus Christ at this time, to which the voice responded that her fervent belief was the reason she had been so chosen.

Betty's encounter with this being is described as being "a profound religious experience." Raymond Fowler comments, "I have seen men with no strong religious background come to the point of tears when listening to a playback of the tape recording of this particular aspect of *The Andreasson Affair.*"

As Betty was returned to her rural home where her family was tranquilly unconscious and unaware of her disappearance, Quazgaa, her escort, gave her parting paternal counsel:

> And he says, "Child, you must forget for awhile." He says my race won't believe me until much time has passed, our time . . . . They say they love the human race. And unless man will not accept, he will not be saved. He will not live. All things have been planned. Love is the greatest of all. They do not want to hurt anybody. But, because of great love . . . they cannot let man continue in the footsteps that he is going. *It is better to lose some than all.* They have technology man can use. It is through the spirit but men will not search out that portion. *Man is not made of just flesh and blood.* He keeps telling me of different things. *Of what is going to take place, what is going to happen. They are going to come to the earth. Man is going to fear because of it.* He says that he had had others here and many others have locked within their minds, secrets. And he is locking within my mind certain secrets. And they will be revealed only when the time is right.[18]

During her hypnotic session to elicit this information, Betty "spoke for a time in an unintelligible tongue, interspersed with phrases like 'Signal Base 32,' and 'Star Seeso.'"[19] Whitley Strieber cites the work of a Mr.

---

17 *The Watchers,* p. 339.

18 *Ibid.,* pp. 339-40.

19 *Paranormal Borderlands of Science,* p. 183.

Keane, an expert in the Gaelic language who has produced phonetic renderings of words of the unknown language spoken by Betty. He claims that the words spoken by Betty corresponded nearly exactly with Gaelic equivalents. He renders the following translation of Betty's "alien" language: "The living descendants of the northern peoples are groping in universal darkness. Their mother mourns. A dark occasion forebodes when weakness in high places will revive a high cost of living; an interval of mistakes in high places; an interval fit for distressing events.[20] The name of Betty's "Quazgaa" translates, curiously, as "One of the cross."

Again, Dr. Mack records a couple of times when humans' belief in Jesus Christ was used by the aliens in their contacts. Jerry, the female abductee who was also the mother of the boy who said that while on the UFO he saw "the King and He is God," further related that "she was shown a picture or painting of Jesus in a white robe. Again the beings just wanted to study her reaction to seeing this image . . . ." (Abduction, p. 134.) Jerry's memory of events during this activity faded right after seeing the picture of Jesus. Dr. Mack's other abductee who was tested with regard to his "Jesus reaction," was Peter. Dr. Mack relates the following.

> Peter's first memory was of looking out at night from a condominium where he was living in Hawaii at "this huge ship hovering over the golf course." At the same time three people, "like three Jesus freaks or something," walked into the door of his living room. "I think they're talking to me about God and about religion, and wanting to convert me or something . . . It's them!" Peter exclaimed. "There's a connection between them and the ship. It's like their human form. They've come again. They've come again. I can't fight 'em. I can't fight 'em. . . ." His thoughts returned to the woman at his door in Hawaii who "kept looking at me . . . I wasn't ready to accept God or Jesus or something . . . ." [*Ibid.*, pp. 317-18]

Year 2000

Betty Luca was assured that the beings do not want to hurt anybody, but, out of their sense of great love for humanity, they cannot let man continue in his present course. Man is a *spiritual* being, not just flesh and blood. When they come, man will fear it, but will benefit from their technology as it applies to his spiritual side. How will man be diverted from his present course to one of spiritual technology? Quazgaa explains cryptically, "It is better to lose some than all." Again, the number two billion

20 Whitley Strieber, *Transformation: The Breakthrough,* pp. 46-47.

comes to mind. As discussed in a later chapter, New Agers lament that around the year 2000, approximately two billion humans of lower vibrational rates will be eliminated. After Laura, Kathie Davis's abducted sister, had overcome her difficulties with hypnosis triggered by her UFO encounter, "she was left with one strong thought: that by the year 2000 the world would be totally different than we know it, but it would be only for the young and strong."[21]

Like thousands of other abductees, Dr. Mack found that his patients were receiving information intended to make them, and us, believe that Earth is about to undergo many changes within the next few years. "When the three beings visited Peter in Hawaii they asked him again 'if I wanted to continue.' Although he was not 'ready to remember' at that time, 'there's always been a choice I've made to go forward.' A new choice seemed to be developing now, 'something else,' Peter said, as he recalled being led down a hallway where 'they're gonna show me things.' As he walked down the hallway, Peter was shown pictures on either side. He saw nuclear explosions, sections of Europe and the United States destroyed, 'a lot of people burned, a lot of people upset . . . the human race changing' its 'form' and 'texture.' Peter and the [alien] woman and the thousands of babies they had produced together were 'part of that change,' of the 'repopulation of the earth.' I asked Peter why this repopulation was necessary. 'Because of the destruction of the earth, because of what's happening, what's going to happen.'

"I asked Peter to say more about that. He said that there was 'a battle going on' over who was going to 'get control of the earth' between 'beings from all over the universe,' not only 'the beings I'm associated with.' This 'has been going on' for perhaps two thousand years, he said, but was now 'all coming to a head.' It was not quite the same as 'good and bad,' Peter explained, but rather 'different possibilities of going forward.' The 'whole reproduction process' in which he was participating was playing 'an integral part' in this unfolding. Peter spoke of prophecy, 'revelations,' and people being 'taken up' as in the Biblical raptures. But to him this had 'nothing to do with religion,' for there were 'ships in space for that to happen.'" (*Abduction,* p. 323.)

Although Dr. Mack recognized that Peter's representations parallelled some of the events associated with the "rapture," the taking up of the events associated with the "rapture," the taking up of the "good" to meet

21 *Intruders,* p. 12.

Christ in the air at His second coming, as outlined in the New Testament of the Bible, he says that "Peter objected to the word rapture, preferring to speak 'of the beings coming to help us to the next place of evolution . . . The slate is going to be wiped clean,' he said. 'It's a new millennium of the earth' that is 'supporting a whole other world.' He had experienced a 'bargaining going on' with 'greater forces actually negotiating the future of the planet,' its 'next two thousand years.'" (*Ibid.*, pp. 323-24.)

Later in the hypnosis session Peter complained that a "cold feeling came over him and he experienced a 'fine vibration' in his body. Images began to come of vast Earth changes that were to occur before the year 2002 . . . . Then he felt he was 'down in the ocean' as great shifts occurred in the continental shelf and a great tidal wave engulfed much of the East Coast and the Gulf of Mexico and swept across the South." Soon, "Peter returned to the theme of human evolution and became 'scared' again as he developed a kind of scenario of salvation. 'Not all of us can go forward in the new being,' he said. There would be a sort of 'straining of those individuals that won't fit into the next place.' The change would be quite beautiful, as consciousness opened 'to a much finer and higher vibration.' His fear related to leaving behind 'what I know,' but 'the higher part of me knows that all of those beings, all of those humans that die actually transcend.' The 'next step,' Peter said, was 'the process of getting these people ready to move through the veil when it's time.' (*Ibid.*, pp. 327-28.)

Salvation

Although it may seem that Betty's message has been thoroughly probed, delivered, and analyzed at this point, other lessons are to be taught to mankind. Her next encounter was in 1973 at the age of 36. Her abductors were somewhat smaller than the last, and were clad in silver suits instead of blue with Phoenix emblems. Betty was kidnapped from her bedroom, unable to wake her sleeping husband, screaming: "Go away! Go away! Lord Jesus! Lord Jesus! Make it go away. Whatever it is, Lord Jesus." It did not go away. At the moment Betty's blanket was pulled off of her, she became sedate. She felt a "pinching" sensation as the beings touched her arm. She floated through the house between two aliens and into a light below the UFO. Suddenly, they, were in the UFO.

Betty found the aliens doing something to an East Indian woman, and she attempted to calm her. To Betty's horror, the beings removed two hybrid fetuses from the woman and placed them in a liquid-filled incubator.

After moving beyond the trauma of the recalled event, Betty was asked to describe what occurred shortly thereafter.

> They're standing in—front—of a glass case. And there's another baby there. A fetus. It's very tiny. And its just laying there inside this liquid. And—But its eyes, they've circumcised the lids. They circumcise the eyelids of those babies and their eyes look so strange. [Betty describes the interior of the "nursery"] They're telling me they have to do this. And I'm saying "Why do you have to do such a terrible thing?" And one of them is saying "We have to because as time goes by, mankind will become *sterile.* They will not be able to produce."[22]

According to the beings, this is the entire purpose behind their accelerated visitation and abduction program. Mankind will soon become sterile and unable to multiply. They, therefore, have the duty of preserving the genetic code of the race on their UFOs until after the holocaust that renders mankind sterile. This explanation is flawed, however, as demonstrated in Betty's description of what follows:

> And they're really pleased with, with this little thing because its eyes are big and black when they cut the lids—like theirs. And they said that the *splicing* took good on this one. And they're telling me that mankind gets so upset when they take the *seed.* And, really, the very first part that man and woman, when they came together, was to bring forth—was not for their pleasure, but to *bring forth.* And mankind keeps on spilling the seed of life over and over again. And they cannot understand why man gets so upset when they take the seed.[23]
>
> And they're telling me that they're doing this because the human race will become sterile by the pollution and the bacteria and the terrible things that are on the earth. They're telling me that they have extrapolated and put their protoplasma in the nucleus of the fetus and the paragenetic. [Betty stops and sounds completely frustrated.] I don't understand them. Something like the paragenetic will utilize the tissue and nutrients to—I don't know—transform the creature or something like that. I don't understand what they're saying. They're saying also about man, that he gets so upset and, and in the beginning that it was meant for bringing forth children, not for pleasure. And, they are taking the seeds so that the human *form* will not be lost—That they too are made of the same substance and, that some of the female fetuses don't accept the plasma very well and that they have to—I don't understand what they're saying—something about—I can't understand and I'm just trying to repeat what they're saying . . . . [*Ibid.*]

If they are "taking the seeds so that the human *form* will not be lost," why are they concentrating on creating a hybrid line from their own

---

22 *The Watchers,* pp. 24-25.
23 *Ibid.*, pp. 48-49.

genetic stock mixed with ours? Many cases demonstrate that either small greys or hybrids are supposedly being produced from these genetic experiments—not human babies, although there is some evidence that some human children are being bred to work with the small greys.

Betty next comments that they are now working on the second fetus taken from the woman, and the hypnotist asks, "Do the beings have blood?

> No. They said they utilized the blood and tissue and nutrients that are there and the *form* and the fetus for the growth of the *new creature.* And some females [alien females] just don't accept the protoplasma all together. So, they grow and use them to carry them, to carry other fetuses but they are very weak and cannot be artificially inseminated like humans. [Q. "What happens to the fetus? Do, do they keep it there, or?] The fetuses *become them*—like them. They said they're *Watchers* . . . and they keep seed from man and woman so the human *form* will not be lost.[24]

This last statement is as significant as any in constructing the extraterrestrial message: "The fetuses *become them*—like them." Although obviously significant, we can only guess at the purported meaning of this pronouncement. In its simplest interpretation, the statement indicates that female aliens have come to have difficulty reproducing, and the alien race is propagated by using human women as surrogate mothers—living incubators. But why the hybrid children instead of pure alien? Their response is that "they keep seed from man and woman so the human *form* will not be lost." Because this explanation does not fully answer the question, perhaps a more complex interpretation is in order.

A complex, yet literal interpretation of the statement is: "the highly spiritual nature of the creatures allows the transmigration of the creatures' spirits from the older, weaker bodies to the younger, genetically improved bodies." With this explanation we literally say that "the fetuses become them." Such an explanation could also account for the longevity touted by the creatures. This latter explanation is quite disturbing—almost as disconcerting as the underlying, simple explanation to the entire phenomenon—they conduct their operation *to obtain bodies;* something that they cannot do without our help for those with strong religious roots, this possibility presents a nightmare of Biblical proportions.

As Betty was readied for high velocity travel, she was placed into a simple "standup kind of seat." The aliens apologized for the inconvenience, explaining that this type of craft was not designed for human transport, and

---

24 *The Watchers.,* pp. 48-49.

this particular type of "seat" was all they had available for her. Just before takeoff, the other being left, leaving Betty alone with one small grey. The remaining being communicated telepathically with Betty: "He's very grateful to me. [Q. Why? For what?] For being there. That I helped the lady calm down, he says. It was very beneficial for the fetus." Betty took advantage of this moment to ask some questions. We gain another piece of the message puzzle from the answers.

> Who are they? I asked. I was trying to ask him. He says that they are the *caretakers* of nature and natural forms—*The Watchers.* They love mankind. They love the planet earth and *they have been caring for it and man since man's beginning.* They watch the *spirit* in all things. Man is destroying much of nature. They are curious about the emotions of mankind. [Q. Do they have emotions?] Not like man. [Q. But, didn't he say they *love* the earth?] It is not the same emotion. It is a forever love—constant, continual. And they are the *caretakers* and are responsible. And this is why they have been taking the *form* of man. [Q. How, how long have they been taking the *form* of man?] For hundreds and hundreds of years . . . He's saying that they have *collected the seed of man* male and female. And they have been collecting *every species* and *every gender* of plant for hundreds of years.[25]

They are the Watchers, the Caretakers—they have been caring for the Earth and man since man's beginning.

In a letter to Raymond Fowler dated July 12, 1988, Betty communicated the following information that she had obtained concerning telepathic communications between her and a being that had abducted her from her trailer.

> Although I did not see him during the sessions, I know now that another *being* was left behind in the trailer as a guard against any kind of intrusion when I was taken up.
>
> I was told, through power, they can form illusions right down to movement, heartbeat, and breathing of a person for the sake of cover. If any outsider was to approach the trailer and look in, they would have believed the moving form on the sofa was actually me. The guard would activate the power to change the thought in the intruder's mind, to turn away. An intruder would have thought it was his natural decision and will, as not to disturb me . . . .
>
> What was revealed to me again is their power can control things for miles around to a small local spot. Something can be happening right amongst the busiest activities of a host of people and yet never be seen by some except those the *beings* choose to reveal it to. The *beings'* scanners and minds pick up any and all life forms within the

---

25 *Ibid.,* p. 119.

> immediate area of a target. They said they're keepers of form. They've been entrusted with and are responsible for the care of all natural form since the beginning. They know physically all there is to know about plant, animal, and human life form with the exception of human emotions which often activates the free will to do as it pleases. Emotions make man unpredictable. That's why it is not the immediate vicinity in their control that concerns them, but unexpected intrusion of someone entering the vicinity. Even though they can quickly gain control of the situation, they may be too busy to detect an invasion.

These contentions of ability to control the minds and actions of many people for a radius of several miles are extraordinary. It is interesting that Udo Wartena's visitors were unable to detect his presence, and were caught off guard as he ascended from his 30 foot mine shaft. Evidently, they could not detect lifesigns through that much earth. Furthermore, if these assertions have any validity, this "intruder" difficulty could well explain how Ed Walters was able to escape abduction on the first attempt. Even he wondered if the airplane coming into the UFO's airspace didn't have a part in his narrow escape. Of course, the above statement brings to our attention another "inconsistency" or "inexplicable fact" of the Ed Walters case—other abductees (almost universally) are of the opinion that if the UFOs want to get you, there is nothing you can do to prevent abduction. Ed Walters on the other hand, evidently successfully dodged and avoided UFOs many times. Again, this may only have been a devised pretense of his abductors, as evidenced by the probability that Ed Walters was apparently abducted on two or three occasions of which he was unaware initially. Either way, it is all another example of a great cat-and-mouse game being played by whoever is behind the UFO phenomenon.

Reports of "aliens" coming through the walls of the house to abduct people have become common. Betty Luca asked the beings how they did it and they replied, "By controlled vibrational levels. It is very simple, those structures are very loose." The entire subject of "vibrational levels" is so closely connected with the New Age movement, occult spiritism, and the predicted annihilation of 2 billion humans that we should pay special attention to these assertions. Further evidence of our need to be wary is advanced in Betty's next observation.

> I asked, "What did you mean 'too many eyes and ears watching and listening'?" He answered that "The physical presence of eyes and ears is no concern for we control this easily. But, waves and manifestations of present energy cannot be erased. What is, is

> always there like grooves in the record of time. If the right tool or point is rubbed against hairlike warps and weaves, the recorded energy is artificially materialized. That is why we have to scramble the energy. When you are taken up, an excessive amount of energy will be scattered about us, masking the identity. This mask will blend and fill in any and all gaps and weaves during transition. Stay very still during the extensity of yourself," he said.[26]

Not only do we gain insight into the nature of time (assuming that any truth exists in this statement), but we learn that what the beings are doing is being concealed from those who would look at the record inscribed in the fabric of time. This self-admitted deceit renders the acts of the "aliens" highly suspect. If, as claimed, "They've been entrusted with and are responsible for the care of all natural form since the beginning" (not that this passive contention informs us who has entrusted them), why is it necessary to conceal their benevolent acts from those who may check the record of time?

Speaking of the beings' prior statements about the continuum of time, Fowler comments, "They insisted that our concept of time was *localized* and that time as we understood it did not really exist. The human concept of time was illusory. All is *Now.*" Betty further related that: "The future and the past are the same as today to them—Time to them is not like our time, but they know about our time—They can reverse time—'Time with us is not your time. The place with you is localized. It is not with us. Cannot you see it?'"[27]

Admittedly, the beings' professed concept of time may only be true from a technological point of view (allowing for time travel), or, if they are actually fallen (or other) spirits, these principles may still hold true. It is evident that the beings exist in our physical plane as well as in the "other self" plane simultaneously, as demonstrated during abductees' OBEs. The beings participate in the scene and interact with the OBE abductee, before, during, and after the OBE. This fact gives us much to think about in our analysis of the origin and nature of the beings.

The Message

As we examine the contacts made by the "aliens" with human abductees, we discover that a systematic delivery of a message is occurring. The mes-

---

26 *Ibid.,* pp. 181-83.

sage is delivered in part by the makeup of the abduction experience, and in part by the actual things that the aliens are telling the abductees. As strange as it may seem, the bizarre message we received from the early contactees, which was entirely rejected by the public, is the same message that we are now receiving through legitimate abductees who are not lying about their encounters. We are still hearing that highly advanced "spiritual beings" are coming to bring us "salvation," both physically and spiritually, but the message is more palatable now because of the trappings that accompany it.

Although we discuss the message and its elements in greater depth in the remaining chapters, more importantly, we look beyond the surface of the message, to that portion that the "aliens" don't want us to discover. The Emperor's new clothes, like the aliens and their message, when viewed without the aid of the "new reality," are nothing but phantoms.

---

27 *Ibid.*, p. 209.

9
What Is The New Age?

A Brief History Of The New Age

To the uninitiated person, the term "New Age" may hold little meaning—denominating a form of easy listening jazz music, perhaps, or indicating an "enlightened" viewpoint on complex social issues. The New Age is a movement of astronomical proportions and consequence, however, and it is incumbent upon everyone to understand the message and import of the New Age movement. Why we must gain this understanding and how it relates to the subject of extraterrestrial visitations to Earth becomes apparent as we survey the beliefs and goals of the New Age movement and as we explore the purported purpose and message behind the wave of UFO visitations to Earth.

None of us can forget the decade of the 1960s. It saw, among other things, the birth of the free speech movement with its attendant mass protestation and denunciation of everything of traditional value. Lasting into the 1970s, vocal leaders, entertainers, and educators decried the avarice and savagery of the contemporary "establishment," demanding that love, peace, and global harmony supplant the evil cultural and governmental systems that oppressed the world. Traditional social and religious institutions were condemned as sponsors and advocates of the evil *status quo* and an entire generation of western civilization was in-structed to "tune in, turn on, and drop out." The parts about turning on and dropping out were in reference to drugs and society, respectively—but the tuning-in portion of the mandate is of special interest to us here.

A new way of life was preached during this time of social and cultural revolution. Free expression, free love, and free lunch became the Holy Trinity of a new generation. The greed and guilt of the Picean Age were obsolete—the (precarious) Age of Aquarius was dawning. Its adherents looked to the East for direction. Zen Buddhism, Hinduism, astrology, astral projection, transcendental meditation, Yoga, spiritualism, psychic surgery, globalism, holism, humanism, mysticism, numerology, iridology, and reincarnation became the elements of a new faith—the religion of Self.

Who Makes Up The New Age Movement?

The Aquarian Age of the 1970s went "main stream" in the 1980s, allowing traditionalists an opportunity to clean up after the "party" of the prior decades. The Aquarians did not die or fade away during this resurgence of conservatitism,—they simply blended into the traditionalist landscape. They became the leaders of political movements and parties, education, and the media. They are comfortable wearing silk ties and pinstripes or blue jeans. They drive European sports cars and monitor the airwaves with Japanese electronics—there are no easily distinguishable delineations of wealth or preference. What the Aquarians have in common is their belief in, and devotion to, the New Age.

What Do New Agers Believe?

The New Age movement is a coalition of distinct, yet compatible, belief systems centered in the task of taking humankind from what it perceives is its current tier of evolution to the next. The next level of human evolution, according to New Agers, is not the imperceptible next rung of advancement that one would expect of natural evolution—it is a quantum leap in human ability and enlightenment. The seeming incongruity of this belief by those who so whole-heartedly embrace theories of natural evolution is resolved by an understanding that humankind will be catapulted into this next evolutionary phase through external assistance. The espoused source and means of this external aid constitutes the delineating distinctions among the various New Age groups,[1] but certain adhesive constants are shared by them all.

New Agers accept man[2] as a spiritual being whose spirit is self-existent[3] and self-realizing.[4] They believe that Man's spirit is individually progres-

---

1 The term "cults" would be more suitable in many instances.

2 The term "man" is, of course, used in a gender-neutral manner. This practice is generally unacceptable to New Agers because it fails to pay homage to their feminist branch, which desires to return to the glory days of goddess worship and matriarchal societies.

3 New Agers perceive themselves individually as "I am," or "I am that I am."

4 Self-realization is the journey and the end, culminating in personal deification; "god-realization."

sing through several successive episodes of mortality[5] but is simultaneously a component of a universal psychic entity. According to New Agers, there are those, on this planet and others, who are currently at low levels of spiritual evolution, and those who are at higher levels. Some have acquired *very* high levels of spiritual attainment, and an elite handful have reached the very zenith of man's spiritual quest. They believe that in order to progress spiritually, we must look within ourselves, not to another being (God). We are our own saviors. Man must get in touch with the true being within him, then call on the abilities of that person to help him gain further enlightenment. Although this is an individual effort and quest, no one can find the way without guidance. Therefore, a spiritual guide (a spirit on a higher level of progress), acts as a guide on one's path to greater spiritual enlightenment. An initiate seeks to attain a state of spiritual openness by clearing his mind of worldly matters and concentrating inwardly. The initiate believes that when he reaches a point of "accommodation," wherein he can be "overshadowed" by a guiding spirit, he receives direct spiritual communication, knowledge, and enlightenment from his spirit guide.

New Age Gurus And God

To aid the millions of New Age practitioners in their quest for spiritual progress and enlightenment, a multi-billion-dollar industry has arisen in the last decade. Miscellaneous Human Potential and Human Transformational organizations offering seminars and instructional cassette tapes have helped millions to see the light of New Age techniques and practices.[6] They teach meditation and visualization and they sell pyramids and crystals, complete with instructions on overcoming poor health or non-believers whose bad vibrations interfere with personal progress.

Although concepts like God and Jesus are often spoken of in New Age literature and practice, it is the universal consciousness or psychic organ

---

5 Most New Agers believe that this progressive course is common reincarnation, the transmigration of one's spirit from a dying body to a birthing body.

6 Many *Fortune 500* corporations have employed New Age firms to teach their techniques to employees. It is thought by New Age watchdogs that New Age "moles" have been planted in corporate human resource departments to proselytize converts by inviting New Age firms in to teach stress reduction and self-awareness clinics.

to which New Agers refer when speaking of God. It is of their own individual spirits that New Agers refer to as God—they do not accept God, our Heavenly Father, as their God. Jesus, although accepted as an individual human, was not *the* Son of God. He was an important spiritual figure, however, an Ascended Master of spiritual enlightenment, teaching global peace and inner spiritualism. New Age gurus teach that Jesus spent a number of years in the East gaining spiritual enlightenment and mastering spiritual practices before beginning his ministry as recorded in the New Testament. They teach that He was not *the* Christ nor was he any more deified than anyone else of his advanced spiritual level. When he died, he reported to the more advanced Ones before continuing his journey in another body in another life. "Christ," to New Agers, is the Christ Consciousness that we each are, unrealized until we attain our Ultimate Consciousness.

The Christ "office" was held temporarily by Jesus of Nazareth according to New Age leaders, but no longer. It is now filled by those who are able to embody the consciousness of the true Messiah through spiritual enlightenment. The new Master, often referred to with names such as the One or the Universal Mind, directs the great global work of transforming humanity into a more highly evolved being. To reach "godrealization" or this ultimate consciousness sought by New Agers, gurus worldwide offer instruction and guidance. For example, one New Age yoga instructor offers a course entitled "Contacting the Tree of Life and the Tree of Knowledge." His brochure claims, "While meditating with another person, you die and enter the Garden of Eden. You see the Tree of Life and the Tree of Knowledge, surrendering to the great evolutionary intelligence to give you what you need."[7]

When And How Will The New Age Arrive?

The New Age message is that this great evolutionary change will occur in the year 2000,[8] facilitating a "quantum leap in elevated brain power"

---

7 Bob Larson, *Straight Answers on the New Age,* p. 46.

8 Interestingly, New Age idol Nostradamus, born in 1503, predicted in his work *Centuries:* "In the year 1999 and 7 months, there will come from heaven the great king of terror, to raise again the great king of the Mongols, before and after Mars shall reign at will." New Agers see this as an opportunity for "The Plan" of the New Age to save mankind, while occult observers see it as the One's attempt to put The Plan into effect.

that will "result in an upward alteration in mankind's vibrational rate."[9] As Bob Larson, one New Age observer, explains:

> Many New Agers refer to our day as the Aquarian era—a time when a mass visitation of angels and Ascended Masters is occurring. Incidental intervention of higher beings in the past has become an invasion of elevated energies. Our brothers in the beyond want only to lead us to unlimited freedom and joy. If we heed the call, we can avoid annihilation and experience the "playground of existence," guided by the "life-force" of the universe.[10]

Of course, even New Agers realize that not all people are "spiritually attuned" enough to make the evolutionary quantum leap. This group is made up of what they refer to as Millennialists—those who hold tenaciously to the "outdated" Jesus the Christ/Messiah belief, who believe that He will usher in the Millennium as foretold in the Bible. In the New Age point of view, those persons with such lower "vibratory rates" will not escape the great annihilation, and approximately two billion of Earth's children will be "eliminated" during the ushering in of this global harmonic period.[11] New Age guru Ruth Montgomery is told by her spirit Guides that the cleansing will be precipitated by "the coming shift of the earth on its axis, which they say will occur near the close of this century, after a devastating war."[12] Montgomery explains that UFOs will play a part in preserving the "enlightened ones" to repopulate the Earth with good seed after the coming catastrophes. "Although most Earthlings will lose their physical lives when the earth shifts on its axis at the close of this century, a good number of enlightened ones will be evacuated by the galactic fleets and returned to Earth for its rehabilitation."[13]

New Ager Barbara Marx Hubbard, in an unpublished work titled *New Age Commentary on the Book of Revelation,* terms those of a lower vibra-

---

9 *Straight Answers on the New Age,* p. 119. Speaking of the specific eating habits of some New Agers, Bob Larson quotes "one macrobiotic proponent" as explaining: "Planet Earth is surrounded by and immersed in a vibrational body of energy, which is conscious." According to this New Age culinary guru, proper diet places one in harmony with this "etheric web of consciousness." At 83, *citing* "Michio Kushi's New Deal," *East West Journal,* January 1976, p. 22.

10 *Ibid.*, pp. 104-05.

11 John Randolph Price, *Practical Spirituality,* (Austin, Texas: Quartus Books, 1985), pp. 18-19.

12 Ruth Montgomery, *Strangers Among Us,* p. 15.

13 *Strangers Among Us,* p. 43.

tional rate "the bad seed." Hubbard's spirit contact explains that the bad seed must be eliminated before the "paradigm shift" can occur. The paradigm shift is something of a code name for the coming shift in world-view, or the quantum leap in human evolution. More benevolent New Age writers explain that this purging will occur in a natural way—that a series of cataclysms will scourge the Earth, eliminating most of the unready. In the New Age view, these natural purgings will send the unprepared on to other lives wherein they may raise their innate vibrational levels and participate in the New Age in another life. However, one less generous spirit contact declares:

> We have proven that all old religions are based on falsehood. Man is deity! Man is divine. When you bow and worship me you are worshipping the essential deity of all mankind. All who oppose this new unity are a cancer in the flesh of humanity and must be put to death for the greater good of all who remain.[14]

The supplanting of the *status quo* with the age of enlightenment is to be implemented through "The Plan." The Plan, according to New Age watchdogs, is a global secret combination led by powerful persons[15] dedicated to Antichrist and his world reign, who are carefully orchestrating global implementation of New Age dogma and practices. The Plan holds no place for the traditional family, free enterprise, nationalism, or Christianity—all relics of the Picean Age.

The New Age Movement Is Pervasive

The New Age movement is not limited to those who indulge in Ouija boards and crystal balls. All of those who believe that the planets have an influence on their lives and daily examine their horoscopes, a large portion of the American population,[16] make up a branch of the New Age. Of course, most New Age participants are unaware of the ultimate designs of

---

14 David Allen Lewis & Robert Shreckhise, *UFO: End-Time Delusion,* New Leaf Press, p. 172, *citing Tribulation.*

15 Not only are these people said to be wealthy and politically well-connected, but are self proclaimed to be of superior makeup—mentally and spiritually. They are worshipped as "supermen" by the New Age elite insiders who are aware of the full import of The Plan. This belief in a race of supermen that will lead the world into the Aquarian Age is of great importance in light of the New Age message of UFOs, as later discussed.

16 More than 50 million people according to a recent Gallup poll.

the New Age elite. It would be truly paranoid to think that hundreds of millions of people are "in" on a global plot to usher in Antichrist. However, hundreds of millions do follow people whose designs are to facilitate the New Age advent of a "new race" of humanity. Only the New Age elite insiders know who is at the helm. Although lower level participants have little knowledge of The Plan, they enthusiastically proselytize new converts to the movement. Tarot readings and dial-a-psychic services have proliferated exponentially, while trance channeling has empowered New Agers with the ability to reach out and touch *everyone.*[17]

Most Eastern religions already fit nicely into the New Age movement, they being the source of most of its beliefs and practices, and their leaders are often its leaders. Satanic cults, much more widespread than commonly thought, have moved from their traditional South American stronghold into North America with surprising speed and potency. Because New Age ideologies are generally intertwined with references to God or Jesus, or at least the Supreme Creator, many liberal Protestant churches or their members have been enticed into the fold.

The New Age pervades Western culture nearly as potently as its Eastern counterpart has dominated that hemisphere's religious sects for millennia. From film and television stars to top political leaders, the New Age receives widespread endorsements daily around the world. The Plan is progressing as planned, and the One, or the Universal Mind, or whatever name is given to the implementing persona of the New Age movement, is poised to step in as soon as the opportunity presents itself. According to New Age beliefs, the opportunity will come in the form of a global cata-

---

17 Trance-channelling, acting as a medium for a disembodied spirit, now allows each New Ager to speak with anyone, anywhere, to gain the spiritual enlightenment sought. Although famous spiritual leaders of the past (and future) are always popular contactees through channelers, extraterrestrials, Atlanteans, and ocean mammals are chic. Approximately 15% of Americans in 1987 believed in trance mediums according to a *USA Today* poll. (Monica Collins, "Not Some Spaced-Out California Concept," *USA Today,* 16 January 1987, p. 1A.) Channelers have received wide recognition by the media since 1987, and anyone with a telephone can be in contact with a dolphin from a distant planet in only a minute for a channeling fee of $25 to $100. A University of Chicago poll shows that 67% of Americans believe they have had a psychic experience. (Art Levine, "Mystics on Main Street," *U.S. News & World Report,* 9 February 1987, vol. 102, no. 5, p. 67.) A Gallup poll of students 16 to 18 years old showed that they too are believers: 51% believe in astrology; 62% in ESP; 24% in witchcraft; and 21% in ghosts.

clysm either natural or manmade.[18] At that time, only the One and his New Age elite will be able to offer mankind the helping hand required to get us through the colossal difficulties imposed by the catastrophe. Those who resist will be adjudged incompatible, even obstructive, and will have to be eliminated for the good of the whole, because in the New Age, "All is One, and One is All." There is no room for dissension—that would not be harmonious.

Earth Spirit

Many New Agers believe that the Earth is a living being—the goddess Gaia.[19] According to them, Gaia is communicating with "Ascended Masters of the Hierarchy of the universe."[20] "They believe that soon our 'space brothers' will raise a human leader from our midst whom they will endow with supernormal powers and wisdom. This man will lead the world to global government and world peace." Many at this time will be imbued with super-human, paranormal abilities for the purpose of facilitating the ushering in of the Aquarian era. Those of a more cultic bent believe that Gaia will be impregnated by the Sun, giving birth to the superhuman manchild. This superhuman leader will help us solve all of our worldly problems—hunger, poverty, crime, pollution, war, and nuclear and industrial contamination. He will lead the world from its present course of self destruction and from the ecological and economic siege it will soon be under, into the New Age, the Aquarian Age of peace and prosperity. But first, the Earth will experience a "cleansing" of those who are incompatible with the New Age, in which cleansing Gaia will willingly cooperate.

Is the New Age and its goals something that we should take seriously? New Age author Brad Steiger quotes Constable, a New Age spirit guide, as stating:

---

18 Trance channeling and other forms of participation in the occult traditionally peak during times of crisis and uncertainty.

19 It is also the belief of many Christians that the Earth is indeed a living creature of God, possessing a spirit. The "Green Goddess" belief of the New Age, however, holds that the Earth is more than a living creature, it is a goddess—having power, being self-willed, and is anxious to facilitate the transformation of humans into the New Age.

20 *UFO: End-Time Delusion,* p. 16.

> Man will win or lose the battle for Earth itself, for he is at once the goal of the battle and the battleground . . . . The stakes in this battle are not the territory, commercial advantages, or political leverage of ordinary wars, but the mind, soul, and destiny of man.[21]

A survey of modern television programming and motion picture themes demonstrates that New Age dogma is indeed saturating our culture. Morbidity is fast becoming a favorite theme—vampires, a current craze, are said to fulfill our spiritual, sexual, and violence needs all at once.[22] *The Psychology Today* article that speaks to this subject bills its article, "Pop Culture's Occult Boom: The Sudden, Curious Allure of Vampires." The article was published in 1989 and things have worsened since then. Motion picture release schedules are to be filled with death, occult violence, and the fulfillment of abnormal passions and appetites. Ready or not, the New Age is coming.

A UFO/New Age Connection

It is not too difficult to perceive a very close tie between the apparent message of the space brothers, as revealed through early contactees and channelers, and the message of the New Age movement. At this point, the New Age is so inextricably intertwined with the UFO phenomenon that it is impossible to distinguish where one ends and the other begins. Shirley MacLaine, the actress who has become a well-known New Age spokesperson predicted, as did spiritual guides, that during the filming of the television movie based on her book *Out On a Limb,* UFOs would appear. They failed to materialize, leaving MacLaine and the Masters of the Universe to explain that the failure resulted from the low vibrational rates of nonbelievers associated with the project.

For two decades a melding of the New Age, UFOlogy, and western culture has been occurring—even our pop music has been lighting the galactic way to help our youth discover for themselves how to be ready when we are finally contacted and enlightened, as directed in a popular song that instructs:

---

21 *The UFO Abductors,* p. 212.

22 *Psychology Today,* "Hunger for the Marvelous: The Vampire Craze in the Computer Age," Katherine Ramsland, November 1989.

With your mind you have ability
you know
To transmit thought messages
through the vast unknown,
Just close your eyes, and concentrate,
that's the way, you see,
Upon this recitation we're about to sing

(Chorus)
Calling occupants, of interplanetary craft
Calling occupants, of interplanetary craft
Calling occupants, of interplanetary,
most extraordinary craft

(Alien Response)
We've been observing your Earth
And one day we'll make contact with you,
We are your friends.

10
The Western View of the New Age

The prior chapter presented an overview of what New Age writers and critics alike consider to be the salient tenets and goals of the New Age movement. The reader will certainly recognize the disturbing similarities between these New Age beliefs and aims and the pre-Second Coming prophecies concerning Antichrist's conquest and rule over the Earth as believed by Western religions. Because of the exceptional subtlety and cunning with which many of these premillennial events and operations will be implemented, according to Western philosophers, many will be deceived by the predicted Antichrist.[1] Therefore, some discussion regarding God's plan for mankind on the Earth as found in Western belief, and moreover in the Bible, is appropriate before continuing with the New Age/UFO connection.

God's Plan

Although the Judeo-Christian tradition and belief concerning God's plan for the development and eventual salvation of mankind has a common genesis, namely the Old Testament of the Bible, unfortunately, we find that for millennia, each sect and denomination therein has fostered and defended a viewpoint somewhat dissimilar from one to the next. This circumstance has arisen through millennia of social and cultural divergence, multiple translations, political expediency, and other conditions that tend to promote disharmony of interpretation and belief.

Recognizing the potential mine-field that must be navigated in establishing commonality of Judeo-Christian beliefs about the purpose

1 "For there shall arise false Christs, and false prophets, and shall shew great signs and wonders; insomuch that, if it were possible, they shall deceive the very elect" (Matthew 24:24).

and paradigm of Earth life, let us step gingerly and lightly as we establish the elements of that commonality. The author apologizes in advance for offenses to individuals' particular divergences from our effort.

Earth Life

Concerning the human sojourn on Earth, first, it is the Judeo-Christian belief that God created the Earth for the habitation of mankind (Genesis 1:1-25). Next, God created man and woman in His general likeness (Genesis 1:26-30), having formed a physical body from the physical elements and placed a spirit[2] into each body, giving birth to a soul—a living individual being (Genesis 2:7). At a later point, death entered the world, at which time the physical body returned to the "dust" of elements from which it was formed (Genesis 3:19), and the spirit returned to God (Ecclesiastes 12:7). This, in a nut shell, is the sum total of a human's existence on Earth. However, there are other indicators that there is a state of being beyond just Earth life, and it is very much connected with our lives on Earth.

An Evil Influence

When God first placed humans on the Earth, the scripture tells us that another being was here. Chapter three of Genesis contains information that a nemesis of God entered into the body of a serpent, and persuaded the humans to transgress a commandment of God. John the Revelator, the Apostle of Jesus, identified the serpent as "the Devil, and Satan, which deceiveth the whole world" (Revelation 12:9). In fact, in these passages, John describes a great war in heaven in which Satan "and his angels were cast out" of heaven "into the earth" (Revelation 12:7-9). Verse four reveals that the number of the angels he took with him was "the third part" of the heavenly hosts.

Jesus also describes the event as His Seventy special disciples reported back to Him that the power He had given them gave them power over evil spirits, saying, "I beheld Satan as lightning fall from heaven" (Luke 10:18). The prominent Old Testament prophet, Isaiah, likewise reported Satan's fall: "How art thou fallen from heaven, O Lucifer, son of the morning"

---

2 It is believed by many that God "fathers" the spirits of mankind, rather than just manufactures them. *See, e.g.,* Acts 17:28 and Hebrews 12:9.

(Isaiah 14:12) calling him by his pre-fall appellation, Lucifer, which means literally, the *Shining One,* or *Lightbringer.* These indications of former greatness in Godly realms are borne out through the Judeo-Christian literature, ancient and modern. Satan, the Devil, was at one time, before the creation of this Earth, a great and noble being, possibly even an archangel. Why his fall? Isaiah responds, "For thou hast said in thine heart, I will ascend into heaven, I will exalt my throne above the stars of God: I will sit also upon the mount of the congregation, in the sides of the north: I will ascend above the heights of the clouds; I will be like the most High" (Isaiah 14:13-14). Satan rebelled and sought the honor of God by attempting to supplant God, but was banished to the Earth for his efforts.

Humans now find themselves sharing a planet with a fallen race of spiritual beings. These beings are numerous, constituting a third of the hosts of heaven. They seek to possess and infest human bodies as demonstrated repeatedly in the scriptures.[3] These fallen demonic spirits are very powerful and attempt to destroy the lives of those whom they invade. It required the priesthood power possessed by Jesus, which He shared with His twelve Apostles and the Seventy, to cast them out of the bodies of their victims (Matthew 10:1). The New Testament is filled with cases of Jesus and His Apostles casting demonic spirits out of the bodies of afflicted persons. Some of these fallen spirits were almost too strong for even the Apostles to cast out, and Jesus explained that casting them out required the additional effort of "fasting and prayer" (Matthew 17:14-21).

The spirits also seek to deceive mankind, as is also amply manifested in the scriptures[4]. Satan himself can transform himself into an "angel of light" for the purpose of working his deceit (2 Corinthians 11:14). In fact, according to the Bible and numerous prophets of God, these beings have been busily engaged in these activities for thousands of years. They have as their ultimate goal the conquest of the planet and the overthrow of God. This they plan to accomplish by bringing their Master to power through deception and unfair advantage.

In reality, every false teaching and incorrect precept in the world is a product of Satan's attempt to lead men astray. He does not tell big lies to lead us to error, or not initially anyway—they are too easily detected.

---

3 *See, e.g.,* Matthew 4:24, 8:28, 9:32, and 12:22.

4 *See, e.g.* those passages from Revelation 13:11-15, 16:13-14, cited below.

Satan takes truth and changes it just enough to convey an untruth. Why does he carry on this incessant campaign of deception? His entire design is to trick humanity into worshipping him as the Universal Master, thereby owning their souls.

According to the Book of Revelation and other prophetic books of the Bible, the pre-Second Coming deception of Satan will be successful for a time. Satan

will indeed take the reigns of world government and religion because precious few will be capable of discerning his methods or withstanding his power. The Apostle Paul warned: "*Even him,* whose coming is after the working of Satan with all power and signs and lying wonders, And with all deceivableness of unrighteousness in them that perish; because they received not the love of the truth, that they might be saved. And for this cause God shall send them strong delusion, that they should believe a lie" (2 Thessalonians 2:9-11). This is a strong indictment against those who will ignore the plain truth, and will join in the new reality, admiring the great beauty of the Emperor's new clothes. To counter the premillennial outpouring of the Lord's Spirit upon His people as prophesied by the prophet Joel,[5] Satan will counterfeit the phenomenon with an escalation of his own spiritualism to show the world that he is indeed the god of this world.

> And I beheld another beast coming up out of the earth; and he had two horns like a lamb, and he spake as a dragon. And he exerciseth all the power of the first beast before him, and causeth the earth and them which dwell therein to worship the first beast, whose deadly wound was healed. And he doeth great wonders, so that he maketh fire come down from heaven on the earth in the sight of men, And deceiveth them that dwell on the earth by *the means of* those miracles which he had power to do in the sight of the beast; saying to them that dwell on the earth, that they should make an image to the beast, which had the wound by a sword, and did live. And he had power to give life unto the image of the beast, that the image of the beast should both speak, and cause that as many as would not worship the image of the beast should be killed (Revelation 13:11-15).
>
> And I saw three unclean spirits like frogs *come* out of the mouth of the dragon, and out of the mouth of the beast, and out of the mouth of the false prophet. For they are the spirits of devils, working miracles, *which* go forth unto the kings of the earth and of the whole world, to gather them to the battle of that great day of God Almighty (Revelation 16:13-14).

---

5 "And it shall come to pass afterward, *that* I will pour out my spirit upon all flesh; and your sons and your daughters shall prophesy, your old men shall dream dreams, your young men shall see visions: And also upon the servants upon the handmaids in those days will I pour out my spirit" (Joel 2:28-29).

This is the magnitude of the deceptive efforts that these beings will bring to bear against Earth's human population, and according to Paul, even "God shall send them strong delusion, that they should believe a lie." What is the great last days lie that will be so enthusiastically embraced by the world, and bring it into Satanic enslavement? We explore the question below.

Life After Life

Most of the Judeo-Christian world agrees that life continues following this mortal existence, in some form or another. The specifics of the nature of that life, and the mechanics of moving from this world to the ultimate place of residence, is probably the source of greatest sectarian disharmony. However, most agree that upon death, the spirit leaves the body and goes to a place reserved for the spirits of the dead. This place is sometimes called paradise,[6] and sometimes called prison,[7] depending, most probably, upon the spiritual state of the person so assigned. Many believe that it is all the same place, but separated by a division of some kind, which serves to separate those who have done good from those who have done evil—a sort of a pre-judgment day division.[8] It is also believed by many that this spirit world occupies our own Earth's physical space but on a different plane of existence.

The next Judeo-Christian belief about the afterlife is that at some point a reuniting of the spirit and body occurs. This is called the resurrection. It is with this tenet that viewpoints diverge immensely. However, prophets of the Old Testament taught the principle[9] and it is abundantly taught

---

6 Jesus said to a fellow executee, "To day shalt thou be with me in paradise" (Luke 23:43). We learn that during the following three days Jesus had "preached to the spirits in prison" (1 Peter 3:19), and as of His post-resurrection encounter with Mary Magdalene (the first to see him), had "not yet ascended to [His] Father" (John 20:17).

7 Isaiah 24:22, 49:9, 61:1; Luke 4:18; and John 5:25.

8 *See e.g.*, Jesus's parable of Lazarus and the rich man, wherein He reveals, "between us and you there is a gulf fixed: so that they which would pass from hence to you cannot; neither can they pass to us, that *would come* from thence (Luke 16:19:31).

9 *See e.g.*, 1 Samuel 2:6; Job 14:14, 19:26 (And though after my skin worms destroy this body, yet in my flesh shall I see god.); Isaiah 26:19 (Thy dead men shall live, together with my dead body shall they arise. Awake and sing, ye that dwell in dust: for thy dew is as the dew of herbs, and the earth shall cast out the dead.); Ezekiel 37:12; Daniel 12:2; and Hosea 13:14.

throughout the New Testament, Jesus himself having been resurrected, accompanied by many of the righteous.[10] Following the resurrection and judgment, persons are assigned to eternal places of glory (or non-glory), as diverse as the varying beliefs about them.

Reincarnation

The doctrine of reincarnation constitutes the very foundation of New Age philosophy. Reincarnation is the belief in the transmigration of the spirit from one body to another in successive life-episodes. This belief is rejected by the Western Judeo-Christian world because it does not fit anywhere into "God's Plan," as discussed above. The Judeo-Christian view is that a single spirit is placed in a body, fills its life, dies, awaits the resurrection, is resurrected, and moves on to eternal realms. Reincarnation provides for a spirit to wander through lives once a man, next a woman, and next a dog, or insect, or extraterrestrial.

Many are "given" memories of their past life experiences, recalling details from these lives in the personae or from the viewpoint of the deceased person. Sometimes, the information seems overwhelming, packed with the emotions and thoughts of the first person observer. The details of information obtained from these past life memories are oftentimes confirmed by subsequent investigation. So then, are elaborate hoaxes being orchestrated by persons who claim such memories for the notoriety of being featured on a prime-time investigation program or are people actually recalling buried memories from past lives? As with UFO sightings and related phenomena, the percipients appear to come from all walks of life, and some are very credible. Whether their perceptions are correct can be questioned, but for the most part, it is clear that these people are telling what they believe to be the truth.

Dr. John Mack discovered in his research that nearly all of his alien abductees were given past life memories from their abductors, and searched for possible explanations for the phenomenon. He speculates, "It is not necessary to postulate that a past life identity belongs literally to the individual abductee in the same way that our bodies are only ours while we

---

10 "And the graves were opened; and many bodies of the saints which slept arose, And came out of the graves after his resurrection, and went into the holy city, and appeared unto many" (Matthew 27:52-53).

inhabit them. As biologist Rupert Sheldrake has suggested, it is possible that there exists a kind of eternal collective memory on which we may all draw. One might, Sheldrake suggests, 'tune in to particular people in the past who are now dead, and, through morphic resonance, pick up memories of past lives.' This does not prove, he says, that 'you were the person' (Sheldrake 1992)" (*Abduction,* pp. 419-20). Sheldrake and Mack pursue an interesting, but unconvincing theory in suggesting that each human life force renders a psychic energy, filled with its life experiences, into which some people are capable of tapping, or tuning, or whatever enables access to the swirling mass of "collective memory." A better explanation is found in the material that follows.

Ministering Spirits

New Agers tell us that without the helping hand of more exalted spirits reaching down to guide and enlighten, people cannot progress to higher spiritual planes—the essential element of their spiritual progress. Through meditation, chanting, and focusing, practitioners claim to receive spiritual communications, knowledge, and ultimate truth from ascended spirits. If the direct connection is too difficult initially, beginners seek the aid of channelers who bring the guiding spirit to them, thereby opening the channels of communication. Channelers and practitioners all claim that they are merely "overshadowed" by these Ascended Masters, but deny that they are "possessed" by them.

Practitioners of Eastern religions seek such spiritual enlightenment from these elevated spirits. Gurus and students alike aspire to the mystic knowledge that is imparted at the pinnacle of this spiritual quest. Universal knowledge and visions of light beings and mysterious places are conferred upon the initiates, and they enter into spiritual realms and receive advanced understanding that only the attuned are privileged to perceive.

These pursuits and practices are not entirely foreign to the Western mind. Christians, for example, in their own way seek enlightenment, revelation and other spiritual communication from the Holy Ghost. The Holy Ghost, according to scripture, is a spiritual being, closely associated with God, who can either radiate enlightenment from afar, or "download" information directly by merging with the recipient. The Apostle Paul speaks in terms of "your body is the temple of the Holy Ghost which is in you" (1 Corinthians 6:19). All of this tells us that spirits can enter occupied bodies, and impart, or directly download, spiritual communications. Great visions, delivered in panoramic 3-D technicolor, complete with digital sound and special effects, along with temperature, joy, pain, and other sensory blasters, are delivered by these spirits, good or evil.

Spirits sent from God, which are relatively few in number, deliver truth. Those sent from Satan, which are numerous, are sent to deceive and promote Satan's destructive purposes. The source of these spirits is not always apparent—it is difficult to discern between good and evil spirits if the evil spirits appear as angels of light or in some other acceptable form, and deliver messages that appear to be valid or good. In fact, it is rare that an evil spirit will do anything that is overtly bad. Satan deceives—that is his stock-in-trade. He is a liar and a flatterer. He and his fallen spirits provide channelers and psychics with mountains of information, intended to lead people's lives in new directions. Yes, they tell psychics what to expect, where to go, how to do it, where to look, and what to wager. But there is an overriding destructive purpose to the game, no matter how much correct information is imparted.

To ancient Israel, God commanded that anyone who communed with such spirits was to be avoided by the people of Israel, and was to be "put to death."

> When thou art come into the land which the Lord thy God giveth thee, thou shall not learn to do after the abominations of those nations. There shall not be found among you *any one* that maketh his son or his daughter to pass through the fire, *or* that useth divination, *or* an observer of times, or an enchanter, or a witch. Or a charmer, or a consulter with familiar spirits, or a wizard, or a necromancer.[11] For all that do these things *are* an abomination unto the Lord: and because of these abominations the Lord thy God doth drive them out from before thee (Deuteronomy 18:9-12).
>
> A man also or a woman that hath a familiar spirit, or that is a wizard, shall surely be put to death: they shall stone them with stones: their blood shall be upon them (Leviticus 20:27; Exodus 22:18).

These same prohibitions are found throughout the Old Testament,[12] and those who sought out channelers or mediums for the purpose of seeking

---

11 Each of these forbidden professions is very much a part of the New Age movement, and is professes, to be necessary to bring one to a higher plane of enlightenment and elevated spirituality.

12 Leviticus 19:31, 20:6, 20:27; Deuteronomy 18:11; 2 Kings 21:6, 23:24; Isaiah 8:19.

spiritual guidance were punished at the direction of God or by God himself. King Saul sought out a woman who acted as a channeler and sought counsel from the recently departed prophet Samuel, as we read in 1 Samuel, chapter 28. Samuel rebuked Saul for his act, which was against the law of God. As explained in 1 Chronicles 10:13, Saul was killed by God for his disobedience, "and also for asking counsel of one that had a familiar spirit, to enquire of it."

Satanic spirits hate mankind. Mankind is destined to receive all that these evil spirits have forever lost, and these evil beings seek to destroy our opportunities by leading us away from the appointed way.

An example of the hate that Satanic spirits have for us is illustrated in the following account. A group of nineteenth century Christian missionaries were sleeping in a building in London, in July of 1837. About daybreak, one of the men came to the others calling for help, complaining that he had been tormented by evil spirits, feeling near death from his experiences. As the men came to his aid, they too were stricken with great pain, and one lost consciousness for a few moments. The men prayed for relief, which came, and report that they then viewed a vision that was opened to their minds, which lasted about an hour and a half. While looking in the direction of the wall, "Space opened before us and we saw the devils coming in legions with their leaders, who came within a few feet of us. They came toward us like armies rushing to battle. They appeared to be men of full stature, possessing every form and feature of men in the flesh, who were angry and desperate."

One of the men reported that while the other was unconscious, before the relief came and the lengthy vision began, he was forced to contend with the evil spirits, who wielded what appeared to be weapons, like knives and scythes, and hurled lies and insults at the men. He beat them back, whether physically or spiritually we cannot tell, and the last "imp" to leave turned almost apologetically and reminded the missionary that he had not spoken a word against the man; to which the missionary replied that like his master, this fallen spirit was a liar, from the beginning. The missionary then rebuked the evil personage, in the name of Jesus Christ, who turned and left as the others had.

These are the spirits that answer the call of mediums and trance channelers, who teach men that there is no God and no devil and therefore, no good and no evil. Many of these beings have the power to provide great visions and information about the future and the past. This ability is the probable source of "past life recall." Either these spirits have viewed these scenes as they occurred, or they generate the scenes from scratch, knowing the essential facts about the scenes. Either way, they download their panoramic psychic projections in such a way that the recipients believe that they are actually recalling buried memories, or receiving a true spiritual transmission of some kind. The percipients are not lying—but a great lie is being implanted and is forming and festering. It will continue to grow and spread until it has accomplished its purpose to destroy the traditional faith in God that is the underlying spiritual strength of the Western world.

Although some *may* escape some of the ill-effects of the premillennial black cauldron that appears too strong for reversal, most of the world will be under Satan's power as foretold by the scriptures: "for by [Satan's] sorceries were all nations deceived" (Revelation 18:23). Speaking prospectively of the time when Satan's premillennial kingdom would be overthrown, however, the Apostle John prophesied that Christ would eventually deliver the world from Satan's grasp: "And [Christ] cast [Satan] into the bottomless pit, and shut him up, and set a seal upon him, that he should deceive the nations no more, till the thousand years should be fulfilled: and after that he must be loosed a little season" (Revelation 20:3).

The New Age Will Bring Antichrist to Power

The scriptures speak of these events with great clarity. Numerous are the verses that detail the rise of Satan to power. Explicit are the details outlining the conditions that will enable his rise to world dominion and the state of the people of the Earth during his reign. Although not explored herein, the reader is exhorted to pursue knowledge concerning these late Sixth Seal and early Seventh Seal (preceding the Second Coming) events.[13] So doing, he or she will be better prepared to withstand the tremendous pressures exerted to force all people to participate in Satan's premillennial kingdom.

---

13 In the Revelation, Chapter 6, John witnesses Christ opening the first six seals attached to the book containing the complete history of the 7,000 year temporal existence of this Earth. Each of the seven seals represents important events pertaining to each of the seven thousand-year periods making up the Earth's temporal existence.

The New Age movement appears to be the vehicle, or a preeminent component at least, by which Satan will establish his premillennial kingdom. The "Network," as New Agers call it, is being put into place even now, awaiting a natural or economic, or some other kind of international catastrophe that will render the world's sovereign nations vulnerable to takeover. The scriptures are quite clear that the premillennial difficulties enumerated in the Sixth Seal portion of the Book of Revelation will supply just such a worldwide crisis.[14] When this occurs, family, religion, and life as we know them will become relics of the Picean Age—that outdated time when Jesus was thought to be the only Christ, the Savior of this broken world.

What does all of this have to do with UFOs? Although all is not completely clear, the occupants of the UFOs, whoever they really are, together with their earthly agents, have been sending us a strong message that they may be very much involved with the coming Aquarian Age. These "aliens" are very New Age.

14 We now live at the very end of the sixth period, whose events are contained in the sixth seal of the book, which events have not yet occurred (although some commentators feel that some may have occurred out of sequence). A thorough study of the sixth seal events leads us to understand that a devastating worldwide earthquake (and resulting related cataclysmic events) will trigger many of the premillennial catastrophes outlined therein, and will certainly facilitate the breakdown of current social, political, and economic systems.

11
Paranormal Properties Of UFOs And Aliens

UFOs Are Not What They Appear To Be

As quoted earlier, New Age author Brad Steiger has observed, "A historical survey reveals that reports of strange objects in the skies are laced through documents of the ancient and recent past. Interestingly, the records seem to indicate that UFOs have adapted themselves to the cultural milieu and the technological capacities of the observers."[1] Steiger's commentary correctly establishes not only the longterm status of the UFO phenomenon, but also its ephemeral, transitory nature. UFOs and their occupants are thought by many to be chameleons—apparitions of something else—something capable of producing supernatural effects in our environment and hallucinatory images in our minds. We discuss these properties in this chapter, and analyze them in the remaining chapters.

Concerning the ephemeral, or paranormal nature of UFOs, Dr. Jacques Vallee concludes after 30 years of UFO research: "I believe that a UFO is both a physical entity with mass, inertia, volume, and physical parameters that we can measure, and a window into another reality. Is this why witnesses can give us at the same time a consistent factual narrative and a description of contact with forms of life that fit no acceptable framework? These forms of life, such as the small grey men seen by Kathie, may be real, yet a product of our dreams. Like our dreams, we can look into their hidden meaning, or we can ignore them."[2] This conclusion is significant, and seemingly grandiose for a mystery that most people feel was solved decades ago. Although important and exciting, the extraterrestrial nuts-and-bolts origin of UFOs is widely accepted and offers an explanation that most people have come to accept as "comfortable." Dr. Vallee's con-

---

1 *The UFO Abductors,* p. 212.
2 *Dimensions,* p. 224.

clusion, however, appears to place the enigma into an unfamiliar paradigm, one with which the masses are uncomfortable—for good reason.

Dr. J. Allen Hynek was originally a scientific skeptic working for the American military, and set out to debunk UFO stories. After much exposure to the UFO phenomenon he became a leading advocate of the extraterrestrial spacecraft theory. However, after reviewing facts like those presented herein, he finally came to think that UFOs may have a paranormal or supernatural origin:

> I would have to say that the extraterrestrial theory is a naive one. It's the simplest of all hypotheses, but not a very likely explanation for the phenomenon we have seen manifesting itself for centuries. In Toronto, Canada, not too long ago, I spoke before a group of liberal-thinking scientists who had gathered for a serious discussion on the latest discoveries in the field of parapsychology. The conference was sponsored by the New Horizons Research Foundation that is ably presided over by Dr. George Owen, a former fellow of Trinity College, Cambridge, England. I told these astute men of learning—including a respected Nobel Prize winner in physics—that we should consider the various factors that strongly suggest a linkage, or at least a parallelism, with the occurrences of a paranormal nature. Among the factors that belie the interplanetary theory is the proneness of certain individuals to have repeated UFO experiences. Another peculiarity is the alleged ability of certain UFOs to dematerialize. A plasma is said to envelope the object in many cases. Then the "cloud" becomes more and more opaque until it completely obscures the UFO. Finally the whole cloud vanishes as though going into another dimension . . . . There are quite a few reported instances where two distinctly different UFOs hovering in a clear sky will converge and eventually merge into one object. These are the types of psychic phenomena that are confronting us in the UFO mystery.[3]

Catherine, an abductee being investigated by Dr. Mack, relates the following occurrence inside a UFO. "They came to another room with a sliding door that slid open upwards. The room seemed to transform from a typically sparse spaceship room with tables, curved walls, and perhaps a viewing screen into an ornate executive conference room complete with shag carpeting, mahogany paneling, and a large viewing screen. As Catherine recalled when she reviewed the tape of this session, she had had the impression that 'the more I thought of a corporate executive conference room, the more it appeared,' but when she realized that this was a kind of staging, the conference room images just melted away to reveal the pre-

---

3 J. Allen Hynek interview, *UFO Report Magazine,* August 1976, p. 61.

vious images, and finally the actual room."' (*Abduction,* pp. 170-71.) In another experience, "Catherine found herself in 'a forest . . . I'm confused but it's there. It's in the room, and there's like trees and rocks and dirt and things off to the left. I can see them from where I am . . . How can I be in a forest?' Incredulous, Catherine exclaimed, 'It doesn't make sense!' for although 'there's forest all around' she could still see the curving walls of the ship. 'It wouldn't have fit. It wouldn't work.' After the regression she reflected that she 'looked way off in the distance' and 'could see walls, but it didn't make sense in context.' She said that the forest even smelled like one and contained pine trees. She estimated it was 'high school gym size."' (*Ibid.*, p. 158.) At page 181 Dr. Mack confirms that another abductee of his reported that the UFO seemed "much bigger inside than outside."

The following account helps to illustrate just how far from "explainable" the UFO phenomenon remains.

The Healing Of Dr. "X"

The fast-paced evolution of the paranormal aspects of UFO sightings and encounters was noticeable as early as the 1960s. One case, reported by UFO researcher Dr. Frank B. Salisbury in his book *The Utah UFO Display,*[4] relates just how metaphysical the phenomenon is becoming. This particular encounter was especially intriguing because it was the best—investigated of its era. The witness was a political figure and Ph.D. in his town in the southeast of France, and the physical evidence was well documented as well as overwhelming. The witness, known as Dr. "X" in UFO literature, chose to remain anonymous at the time because of his political standing, but he was personally known to some of the best scientific investigators in UFO research, and was perceived to be completely credible. Salisbury relates the experience as follows:

> During the Algerian War, he was caught in a mine blast (May 13, 1958), and after several months of recuperation was left with a hemiparesis on the whole of his right side. He could not stand on his right foot unassisted, and he lost much of the use of his right arm, making it nearly impossible for him to play the piano, which he had formerly done very well—and, as we shall see, was able to do again.

---

4 *The Utah UFO Display: A Biologist's Report,* pp. 211-18. The full report of the incident is published in the *Flying Saucer Review:* Aime Michel, 1969, Special Issue No. 3, September, "UFO Percipients," *Flying Saucer Review;* and *Flying Saucer Review,* November-December, 1971, No. 17, (6):39.

On October 29th, 1968, as he was chopping a stump on his property, a slip caused the stump to strike the front edge of the left tibia, producing a superficial wound with a broken vein that led to enough swelling to deform the trouser leg and to cause intense pain. In spite of antibiotics and drugs administered to reduce the inflammation, the wound was extremely painful and the swelling extensive. This was on the night of November first just before the sighting.

Some time before four in the morning of November 2nd, the doctor's fourteen-month-old child calls out. He is expressing the onomatopoeic sounds by which he designates everything shining—a sort of *rho! rho!* The doctor arises (his wife remains asleep), fills the empty feeding bottle, and gives it to the baby. The baby is pointing at the window where periodic flashes of light are showing around the edges of the shutters. The doctor is preoccupied with the pain in his leg and pays little attention to this phenomenon. He hears a shutter banging on an upstairs window and, after painfully making his way up the stairs, gets his first look at the countryside. It is being illuminated by flashes of pale light about the color and intensity of the full moon. The wind is blowing and a storm is brewing, so he at first thinks of lightning, but the flashes are coming at intervals of about a second in a manner quite atypical of lightning. He cannot see the source of the light from this window, so he returns to the kitchen, pours himself half a glass of cold water from the refrigerator (indicating how low his level of excitement remains at this point!), and then makes his way onto the terrace. The clock in the kitchen reads 3:55 a.m.

As he steps onto the terrace, he finally sees the source of the flashing light: it is coming from two luminous objects hovering at some distance to his right. The objects have the double convex form reported by several Uintah Basin witnesses. The top half is a luminous silvery white, but not as brilliant as the full moon; the bottom half is a deep sunset red, brighter at the top than at the bottom. Two "antennas" extend out horizontally, one from each edge of each object. A third antenna on each object is located at top center and is perfectly vertical . . . .

A vertical, cylindrical, white shaft of light is coming from the lower center point of each object. It illuminates the thin mist hanging in the valley below his home. As soon as the doctor steps on to his terrace and notices the objects, they begin to move to the left and toward him. In a moment or two, the points where the two light beams strike the ground can be seen at the top of a small hill. From then on as the objects flash on and off with their one-second periodicity, illuminating the surrounding country side (up to a distance of a few kilometers), with the weak white light, the doctor is forming almost photographic images in his memory. Later, when Michel[5] investigates, the doctor can draw the objects on photographs taken from his terrace, indicating at each point in time their size and the positions of the beams of light on the ground below. White light is also emanating from around the antennas, and the doctor has the impression that these objects were "sucking in the atmospheric electricity and that I

5 Aime Michel is the chief investigator of this case, and was personally acquainted with Dr. Salisbury. He related the details of the incident as Dr. Salisbury drove him to the Salt Lake airport in early 1972.

could see it entering through the antennas and then exploding between the two objects, the whole thing producing one single glow of light." It is this glow that is lighting up the countryside more than the narrow vertical beams.

The objects continue to make their broad sweep to the left, at the same time approaching each other. As they nearly touch, light seems to jump between the approaching antenna. It is bright, but not so brilliant that the doctor cannot observe it directly (e.g., the way one can observe a fluorescent lamp). When the objects are nearly in front of the house, an amazing thing happens. They continue to come together until they actually touch and then begin to interpenetrate each other. This continues for a moment or two until the interpenetration is complete, and only one object identical in its structure to the previous two is now visible . . . .

Just before the interpenetration, the flashing lights on the antennas cease, and the countryside becomes dark again. (The vertical beams of light illuminate it only for a short distance.) After the two objects become one, this object swings closer to the witness. When it is directly in front of the house (on a line at right angles to the face of the house), it begins to move directly toward the doctor. As it approaches, it appears to grow in size until it is enormous. At its closest approach, the doctor is able to accurately estimate its size in relation to the houses and trees below. Michel later measures the diameter at about sixty-five meters (about two hundred feet, exclusive of the antennas, which were estimated at about seventeen meters each, equal to the thickness of the object at its center).

At this point, the doctor, who was first astonished and then perplexed, begins to experience considerable fear. With the object so close, he is able to observe it in detail. He can see a protrusion on the bottom from which the beam of light is being emitted. The bottom half has sections, and there is a dark line that appears to move from the middle toward the bottom much as interference lines sometimes drift across a television screen. This phenomenon holds him transfixed, so that he is unable to run inside and get his movie or still camera.

The object remains stationary for what seems like a long interval of time, then suddenly the doctor notices the beam of light moving toward him. This is not because the object is moving, but because it is beginning to tip or rotate on an axis through the two protruding antennae, so that the top antenna is moving away from him. Finally, the object must have tipped almost ninety degrees, for the light beam illuminates the doctor and presumably all of the front of his house. He instinctively raises his arms to cover his face.

All of a sudden there is a sort of a "bang," the first sound during the entire encounter. Then, according to the witness, the object dematerializes, leaving nothing behind but its cloudy, whitish, fleecy shape, which at once disintegrates and is borne away eastward by the wind. A very luminous, fine, white thread shoots out vertically in a fraction of a second toward the sky and vanishes there into a small white shining dot, which then vanishes with the sound of an explosion like an aerial bomb on the 4th of July.

The doctor experiences considerable shock. He goes indoors and notes that it is 4:05 a.m. He gets a note pad and writes down the details of his sighting along with sketches

> (thereby qualifying himself as one of the best UFO witnesses on record). He awakens his wife and tells her all that he has seen. Both are deeply moved. Suddenly she cries, "Your leg!" The doctor, who is walking to and fro talking excitedly, has lost his limp. He pulls up his pajama trouser leg, and the wound is healed, both the swelling and the pain completely gone. They do not return. Michel visits five days later and is able to ascertain the facts in relation to the healing of the wound not only from the doctor and his wife, but from others including the doctor's physician.
>
> The doctor falls asleep and remains in a sound sleep until two o'clock the following afternoon (his wife awakens at ten that morning). When he awakens, he remembers nothing of his experience until he suffers a fall on the living room stairs, which produces a bump on his head and the full memory of the previous night's events!
>
> For a few days, the doctor experiences further shock, and this produces certain physical symptoms of illness. During this time, he becomes aware that the symptoms of his Algerian War wound seem to be no longer with him. Finally, as the symptoms of shock disappear, it becomes apparent that this is indeed true. And once again he is able to perform expertly on the piano.

The investigators brought in many experts in various fields to help establish the truth or deceit in the doctor's story. Within a few days, the doctor began experiencing discomfort in the area of his navel. The skin became red, forming a perfect isosceles triangle. The researchers were really stumped when the fourteen-month-old baby developed an identical triangle on his navel. The triangles come and go periodically, in tandem evidently. As a follow up to this account, the investigators relate that the doctor complains of continued paranormal activity in his life following this close encounter. He has experienced telepathic and poltergeist phenomena, as well as levitating. Electrical circuits and appliances are also known to malfunction in the doctor's presence. The investigators believed at the time that these subsequent claims of paranormal activity diminished the doctor's credibility concerning the initial encounter. However, as we examine herein, such paranormal phenomena following close encounters have become common.

UFOs And OBEs

Many commentators are beginning to accept as routine the out-of-body experience phenomenon in connection with UFO encounters. In an effort to demonstrate that Betty Andreasson Luca's UFO/OBE episodes are not as atypical as a newcomer would assume, Raymond Fowler cites a 1973 case in which an Army Reserve helicopter was approached by a UFO, and tractored from 1,700 feet in altitude to 3,800 feet.

Captain (later Lt. Colonel) Lawrence J. Coyne, the helicopter commander, was contacted by the Department of the Army, Surgeon General's Office, and asked if he or any other crew member had experienced any unusual incidents or dreams subsequent to the UFO encounter. The "number one" question concerned dreams of body separation. Coyne reported that he had experienced an out-of-body episode in which he got out of bed during the night, only to find he had left his body in the bed. After a brief moment of amazement, he lay back down where his body was, feeling as though he were sinking into something, and woke up. Additionally, Coyne was quite disturbed by a very vivid second dream in which he heard an authoritative voice say, "The answer is in the circle. He looked down to see he was holding a clear, bluish-white sphere in his hand.[6]

Crewman Sgt. John Healy also reported out-of-body experiences following the UFO encounter, that might be better described as near-death experiences. The Pentagon's interest in these specific areas of OBEs following UFO contact appear to support numerous similar OBE reports by nonmilitary witnesses.

Dr. Mack relates an OBE of Eva, one of his abductees. "The other incident occurred while she was on the night shift in the air control. At a slow time—perhaps three in the morning—she put her head down to doze and then 'saw myself floating from the ceiling . . . My consciousness was up there. My physical body was down there.' A 'voice said, 'Come with me, it's good,' and 'I knew at that point I had a choice of living or dying.' Although her heart was beating fast and 'I was sweating like crazy,' Eva was not aware of any life-threatening illness. She said, 'I wasn't interested in dying,' and I said, 'No, I'm not coming." Eva 'knew' she could have died but does not understand why, and found the episode confusing." (*Abduction,* p. 244.)

---

6 *The Watchers,* pp. 188-90.

UFOs And ESP

Along these same lines, civilian reports indicate that frequently, paranormal activity follows the UFO experience participant. Heightened extra-sensory perception (ESP) and poltergeist-like phenomena are often reported. Uncontrollable levitating and electrical disturbances also occur. Some investigators claim that the U.S. government is (or has been) especially interested in following up on UFO sightings to learn of the existence of such post-experience occurrences.

In many cases, contactees, or even those who only witness UFOs, come away with expanded enlightenment, spiritual and/or scientific. Nearly a full one third of UFO encounters are said to include a psychic experience. This is often manifested as telepathic communication with the UFO occupants. Interestingly, those who claim psychic tendencies are more likely to experience UFO encounters. Not surprisingly, self-proclaimed psychics nearly always claim contact with UFOs and their occupants.

Whitley Strieber, the famed fiction writer whose book *Communion* informed the world that he was plagued by "visitors" closely akin to "aliens," reports that he has a great deal of interest in the occult, including astral projection, a self-induced form of travelling out of one's body. Which comes first, paranormal experiences or interest in paranormal activities? It is hard to tell. One may be inclined to believe that the experiences lead to personal investigation. However, those who feel "compelled" to investigate UFO-related paranormal phenomena often discover that they have latent experiences buried in the subconscious realms of their minds—evidently, anyway.

Strieber, as well as many other contactees, has discovered the presence of ongoing "alien" contact and paranormal experiences throughout his life. He rejects the idea that the "visitors" are extraterrestrial. He feels that they are interdimensional beings, who have visited our "phase" of the world throughout human history, communicating with us through demonstration and theater. Strieber describes having his interest in "ancient nature religions and shamanism" piqued by his childhood memory of sitting at a table and working out an anagram the result of which was, "We work by ancient laws." Whoever the visitors are, Strieber believes that they gave him this anagram, and therein revealed something about themselves to him.

The Entity Enigma

Respected journalist Ed Conroy has published a work that goes behind Strieber's *Communion,* analyzing the data and interviewing witnesses. He writes concerning Strieber's conclusion:

> In actuality, Strieber's line of thinking is reflective of the vast array of experiences connected with and explanations offered for what some investigators have come to call the "entity enigma." At first glance, it would seem that the content of Strieber's story is no different from hundreds of cases on record of people reporting terrifying experiences with "bedroom visitors," ghosts, poltergeists, apparitions, religious visions, and what have been regarded as demons. The entire array of such phenomena is beginning to be studied by serious UFOlogists with training in the psychological and social sciences, most notably by Hillary Evans, whose *Visions, Apparitions, Alien Visitors: A Comparative Study of the Entity Enigma,* makes an attempt to thoroughly survey the colorful panoply of anomalous phenomena concerning strange, human beings.[7]

Many guests at Strieber's upstate New York cabin were exposed to the "visitor" phenomenon, including Ed Conroy. One such weekend guest, Philippe Mora, director of *Death of a Soldier* and *The Howling III: The Marsupials,* was hired to direct the motion picture *Communion.* In his interview with Conroy, Mora related the following:

> Then we all went to bed, and I still believe—I had one of these nightmares, quite vivid dreams. Now, you've got to bear in mind that I had been working with Whitley on the script, and was psyched up, although I can't say when I went to bed I was. I was relaxing. I had the experience of lights blasting through the bedroom window, lights blasting under the crack under the door of the bedroom—I tried to turn the light on in my room, and I couldn't turn it on and I was pushed back into the bed. All the while I was consciously saying to myself, "This is a hell of a nightmare."
>
> Then I remember being outside the guestroom door, in the kitchen area, and the whole cabin lit up—every opening, every exterior opening, the whole thing was lit up with moving lights. And I looked through Andrew's bedroom. The door was open, and he was asleep, lit up. I remember Anne [Mrs. Strieber] rushing up to me and saying, "Whatever you do, don't wake Andrew." And then I woke up the next morning, . . . very scared.

---

7 Ed Conroy, *Report on Communion,* William Morrow Company. Inc., New York (1989), p. 261.

Reality Testing

Like Strieber and many others, Mora relates that the experience occurred while in a semiconscious state. The inability to perceive whether or not the percipient is dreaming the experience or living it is a classic characteristic of such experiences. This is true, again, of both UFO occupant encounters and poltergeist and related paranormal phenomena. Strieber relates that on one occasion when the visitors abducted him, he was being floated helplessly through his house, when he decided to perform a "reality testing" experiment—he grabbed the family cat on the way through.

Upon arriving at the quite plain looking "office" of some rather ordinary human-appearing persons (one woman even had her hair in a bun and glasses on her nose), Strieber was chastised for bringing the cat along. When asked why he had done it he replied that he just wanted the company. After some quiet discussion among the entities, they told him that he had made a serious mistake, but that they would remedy it. With that, one of them touched the cat's leg with a small triangular object, and the cat went limp immediately. The day following the abduction experience the cat slept for most of the day and that night. It slept through most of the days that followed, and did not regain its normal strength and habits for several months afterward.

Historical Parallels

A great deal of research is currently underway comparing the behavior of the UFO phenomenon with that found in historic myth, legend, and fairy tales. Initially, this seems like a long shot, even childish. However, the parallels are striking, and research appears to be bearing enlightening fruit.

Jacques Vallee has compared much of the pertinent literature and shares the following legend as being typical of one type of parallel experience:

> According to the Paiute Indians, California was once populated by a superior civilization, the *Hav-Masuvs.* Among other interesting devices, they used "flying canoes," which were silvery and had wings. They flew in the manner of eagles and made a whirring noise. They were also using a very strange weapon: a small tube that could be held in one hand and would stun their enemies, producing lasting paralysis and a feeling similar to a shower of cactus needles.[8]

---

8 *Dimensions,* p. 86.

These descriptions are common in the folklore of native tribes and ancient civilizations. So, who are these airborne, stun-gun toting tribes that are reported worldwide throughout recorded history? We generally assume that they are mere tribal embellishments of feared enemies. Of course, the fact that they are recorded in similar terms throughout the ancient world tends to lend weight to the veracity of the legends.

The Secret Commonwealth

For hundreds of years European scholars have attempted to document and piece together the legends of nonhuman beings in their region, historical and contemporary. A wonderful synthesis of the phenomenon was authored by Reverend Kirk of Aberfoyle.

> In the last half of the seventeenth century, a Scottish scholar gathered all the accounts he could find about the *Sleaghmaith,* and, in 1691, wrote an amazing manuscript entitled *The Secret Commonwealth of Elves, Fauns, and Fairies.* It was the first systematic attempt to describe the methods and organization of the strange creatures that plagued the farmers of Scotland. The author, Reverend Kirk, of Aberfoyle, studied theology at Saint Andrews and took his degree of professor at Edinburgh. Later he served as minister for the parishes of Balquedder and Aberfoyle and died in 1692.
>
> Kirk invented the name "The Secret Commonwealth" to describe the organization of the elves. It is impossible to quote the entire text of his treatise, but we can summarize his findings about elves and other aerial creatures in the following way:
>
> 1. They have a nature that is intermediate between man and the angels.
>
> 2. Physically, they have very light and fluid bodies, which are comparable to a condensed cloud. They are particularly visible at dusk. They can appear and vanish at will.
>
> 3. Intellectually, they are intelligent and curious.
>
> 4. They have the power to carry away anything they like.
>
> 5. They live inside the earth in caves, which they can reach through any crevice or opening where air passes. . . .
>
> 8. Their chameleon-like bodies allow them to swim through the air with all their household.
>
> 9. They are divided into tribes. Like us, they have children, nurses, marriages, burials, etc., unless they just do this to mock our own customs or to predict terrestrial events.
>
> 10. Their houses are said to be wonderfully large and beautiful, but under most circumstances they are invisible to human eyes. Kirk compares them to enchanted islands. The houses are equipped with lamps that burn forever and fire that needs no fuel.

11. They speak very little. When they do talk among themselves, the language is a kind of whistling sound.

12. Their habits and their language when they talk to humans are similar to those of local people.

13. Their philosophical system is based on the following ideas: nothing dies; all things evolve cyclically in such a way that at every cycle they are renewed and improved. Motion is the universal law.

14. They are said to have a hierarchy of leaders, but they have no visible devotion to God, no religion.

15. They have many pleasant light books, but also serious and complex books dealing with abstract matters.

16. They can be made to appear at will before us through magic.[9]

Following is another example typifying the parallels between legendary abduction accounts and those found in modern UFO literature.

A Swedish book published in 1775 contains a legal statement, solemnly sworn on April 12, 1671, by the husband of a midwife who was taken to fairyland to assist a troll's wife who was giving birth to a child. The author of the statement seems to have been a clergyman. "On the authority of this declaration we are called on to believe that the event recorded actually happened in the year 1660. Peter Rahm alleges that he and his wife were at their farm one evening late when there came a little man, swart of face and clad in grey, who begged the declarant's wife to come and help his wife then in labor. The declarant, seeing that they had to do with a Troll, prayed over his wife, blessed her, and bade her in God's name go with the stranger. She seemed to be borne along by the wind.

In another tale, the midwife's husband accompanies her through the forest. They are guided by the "earthman" the gnome who has requested their help. They go through a moss door, then a wooden door, and later through a door of shining metal. A stairway leads them inside the earth, to a magnificent chamber where the "earthwife" is resting. Kirk reports that in a case whose principals he personally knew the abducted woman found the home of the Little People filled with light, although they could not see any lamp or fire.[10]

Magonia, or fairyland, is thought to be a remote country or invisible island reached after a long journey, aerial or otherwise. Vallee observes, "But a second—and equally widespread—theory, is that Magonia constitutes a sort of parallel universe, which coexists with our own. It is made

9 *Dimensions,* pp. 87-89.

10 *Ibid.,* p. 127.

visible and tangible only to selected people, and the doors that lead to it are tangential points, known only to the elves and a few of their initiates."

Vallee cites Hartland, who offers tales that illustrate the latter theory, such as the following:

> In Nithsdale a fairy rewards the kindness of a young mother, to whom she had committed her baby to suckle, by taking her on a visit to Fairyland. A door opened in a green hillside, disclosing a porch which the nurse and her conductor entered. There the lady dropped three drops of precious dew on the nurse's left eyelid, and they were admitted to a beautiful land watered with meandering rivulets and yellow with corn, where the trees were laden with fruits which dropped honey. The nurse was here presented with magical gifts, and when a green dew had baptized her right eye she was able to behold further wonders. On returning the fairy passed her hand over the woman's eye and restored its natural powers.[11]

The parallels with modern descriptions of UFO occupants and related entities are remarkable. Strieber describes an abduction in which he was made to think he was on a strange airplane. He then relates how the stewardess gave him some eye drops for motion sickness (yes, he noted the strangeness of this remedy), after which he was taken to exotic places.

Vallee cites many cases wherein the historical beings are described as having glowing eyes and many other characteristics identical to those ascribed to many UFO occupants, and Kirk documents humans' ability to make pacts with the beings for personal gain or for protection from them.

Magic Wands

Ed Walters, like many other UFO witnesses, describes certain wandlike instruments carried by the "aliens." These silver wands or rods are apparently the same as that carried by the single small grey on the night it attempted to lure Ed Walters out of the house into the clutches of the blue beam. Although we can assume that such a glowing silver rod was nothing more than a stun gun, or communication device, the presence of the silver rods that "glowed" is worth noting from a traditional, or paranormal viewpoint. Betty Luca, as well as others, describes how such devices are used to manipulate the environment, ambiance, or technological appliances in the vicinity of the encounter.

---

11 *Dimensions,* p. 128.

Certain ancient religious texts occasionally refer to superior beings, or those sent with ecclesiastical authority, as possessing various types of rods, holding them in their hands as though they conveyed authority. For instance, from the *Apocalypse of Abraham* we read: "And I rose up and saw him who had grasped me by the right hand and set me upon my feet: and the appearance of his body was like sapphire, and the look of his countenance like chrysolite, and the hair of his head like snow, etc. and a *golden scepter* was in his right hand. (Section X.)" Also, from the Iliad we read: "'And the Elders were seated upon shining stones [stones of truth] in a holy circle,' with scepters of inspired utterance in their hands . . . ." (Hugh Nibley, *Abraham in Egypt, quoting Iliad,* XVIII, 497-508) Besides these, our scriptures, as well as our secular monarchies, are replete with examples of the possession of such rods or scepters as symbols of authority or power.

Why rods that appear like scepters? Why not, for instance, small handheld spheres or cubes—why always scepters? No other analogy is suggested here except for the universality of the principle.

Crossing The Time Zone

Another characteristic of fairyland that is also common in UFO abduction cases is what the author terms the "Rip Van Winkle effect." When visiting with the elves or fairies or Gentry, humans are said to be missing for days, months, or years, when to them the time passed as though it were only minutes or hours. In fact, it is based on this rich folklore that Washington Irving wrote *Rip Van Winkle.* The reverse is true also—the abductee often feels that he or she was in fairyland for weeks or longer, only to find that no time has passed in our world.

> This is our fourth point, and quite a remarkable one. Time does not pass there as it does here. And we have in such stories the first idea of the relativity of time. How did this idea come to the storytellers ages ago? What inspired them? No one can answer such questions. But it is a fact that the nonsymmetry of the time element between Magonia and our world is present in tales from all countries.[12]

---

12 *Dimensions,* p. 131.

In many UFO abduction cases the victim's absence went unnoticed because no time had passed. In others more time had passed than was realized by the victim. Of course, there exist other plausible explanations for such phenomena—drugs, for example. These are plausible, but not generally probable. In one case, a military watch in South America approached a landed UFO, and one of the men went over the knoll to get a closer look. A few minutes later he was found by his colleagues, disoriented, with several days' growth of beard.

Elves And Fairies

Describing the UFO entities Vallee says:

> Entities human witnesses report to have seen, heard, and touched fall into various biological types . . . . Most of the so-called pilots, however, are dwarfs and fit into two main groups: (1) dark, hairy beings—identical to the gnomes of medieval theory—with small, bright eyes and deep, rugged, "old" voices; and (2) beings—who answer the description of the sylphs of the middle ages or the elves of the fairy-faith—with human complexions, oversized heads, and silvery voices. All of the beings have been described with and without breathing apparatus. Beings of various categories have been reported together. The overwhelming majority are humanoid.

In describing the entities' behavior Vallee adds:

> The entities' reported behavior is as consistently absurd as the appearance of their craft is ludicrous. In numerous instances of verbal communications with them, their assertions have been systematically misleading. This is true for all cases on record, from encounters with the Gentry in the British Isles to conversations with airship engineers during the 1897 midwest flap and discussions with the alleged Martians in Europe, North and South America, and elsewhere. This absurd behavior has had the effect of keeping professional scientists away from the area where the activity was taking place. It has also served to give the saucer myth its religious and mystical overtones.[13]

Vallee feels that the entities and their UFOs, whoever they are, "are paranormal in nature and a modern space age manifestation of a phenomenon that assumes different guises in different historical contexts."[14] In an interview with *Fate Magazine*[15] Vallee said, "We have evidence that the

---

13 *Dimensions,* p. 166.

14 *Ibid.,* p. 213.

15 "Vallee Discusses UFO Control System," p. 65.

phenomenon can create a distortion of the sense of reality or to substitute artificial sensations for the real ones."

Pursuing his paranormal activity theory, Vallee suggests, "It is conceivable that there is one phenomenon that is visual and another that creates the physical traces. What I am saying is that a strange kind of deception may be involved."[16] He also notes, "It seems as if an external force takes control of people. In the close encounters people may lose their ability to move or speak; in the abduction cases, which are the most extreme example, they gradually enter into a series of experiences during which they lose control of their senses."[17] John Keel, a well respected UFO researcher similarly relates that "over and over again, witnesses have told me in hushed tones, 'you know, I don't think that thing I saw was mechanical at all. I got the distinct impression it was alive.'"

Vallee concludes his study of similarities between UFO occupants and the Gentry of the past stating: "The UFO occupants, like the elves of old, are not extraterrestrials. They are the denizens of another reality." What does this tell us? That top scientists in UFO research appear to accept the underlying theory that UFOs and their occupants are nothing more than modern versions of historical manifestations of spirit entities.

Modern Gentry

Raymond Fowler was contacted by a young man who saw a UFO. He said that a "few years later he woke up to see a *beam of light* entering his bedroom. It contained an entity with long blond hair. Like [Fowler, as a child], he could not tell whether it was male or female. Like [Fowler], he was filled with an indescribable feeling of love." These encounters with angelic appearing humans are perceived almost as religious experiences by the witnesses—for good reason—this is clearly the intent of the entities.

Fowler relates that his father had been hit by "lightning" while stationed at a U.S. Naval Radio Compass Station. The lightning came through the equipment and hit him, lodging behind his solar plexus, "where it remained and revolved like a fiery sun" inside of him. He sat baffled wondering why

16 *Fate Magazine, (Ibid).* p. 63.
17 *Ibid.*, p. 64.

he wasn't dead, watching as the eight-inch ball of energy whirled inside him, pulsating to the beat of his heart.

Fowler's father looked up to see a beam of light coming through the station roof, going up to a "radiant star." The light also went down deep into the rock beneath the station. He sat for a long time, feeling outside of time and space. Then the light expanded, from which three entities sprang. He described them as "majestic-looking smiling men in shining robes of light," who communicated with him telepathically. He formed questions concerning the startling events that were occurring, but his visitors did not answer them.

His father further described the three as "fine-featured" and having "light cream-textured complexions. Their eyes were so bright it was difficult to see the color but [he] thought they were blue." He said they wore odd hats that were the same shining blue as their robes. One of the beings pointed at him and the ball of energy leaped from his body into the being's hands. The three of them tossed it between them until it disappeared. The three smiled, bowed, and disappeared into three flashes of light. The Radio equipment was incinerated.

Although there is no direct relationship to UFOs in this account, the fact that the Fowler family was generally involved in UFO sightings is illustrative of the phenomenon—those who have UFO encounters often have other anomalous experiences. It is obvious that this was no common lightning strike incident. From beginning to end the account is better described in spiritual terms.

Many years after Fowler's father had the above experience, he related a very vivid dream he had to Fowler and his wife. He dreamed he went into a large movie theater, sat down, and watched as the lights dimmed, the curtains opened, and a large silver screen was revealed. Trumpets blew as a date appeared on the screen: "October 4, 1957." He heard a loud voice declare, "On this day the Cosmic Age will be ushered in!" The date faded, the curtains closed, and he left the theater. Fowler relates that although he and his wife rolled their eyes at each other in the usual incredulous fashion, he did write the date on a scrap of paper. It was years later when the predicted date arrived—the headlines that morning announced that Sputnik had launched mankind into the Space-Age. Fowler remembered the date and ran to fetch the scrap of paper. The date was the same. It is interesting that Fowler's father made another prediction that came true also—the fascinating part is that it dealt with a UFO sighting.

Catholic UFOs

Throughout history there have been thousands of anomalous encounters with the unknown that have the appearance of religious manifestations. Many of these meetings contain components characteristic of UFO encounters. Many researchers of the UFO phenomenon have noted the well-attended and well-documented case of the "Miracle of Fatima." Dr. Salisbury offers the following condensed account:

> For six months beginning in May of 1917, three children near Fatima, Portugal, were visited on the thirteenth of each month by an apparition that identified itself as the Virgin Mary. The transparent little blue lady with her hands in an attitude of prayer was seen by Lucia de Santos (age ten), Jacinta (age nine), and Francisco (age seven), but she was heard only by the two girls. She gave them messages quite appropriate to Catholics in the Portugal of 1917. Each time she appeared, a larger crowd of people was on hand, but only the children were able to see her. Finally she said that on October 13th, a sign would be given that would convince everybody. It is estimated that some 70,000 people were milling around in a muddy field during a light rain storm waiting for the sign. Suddenly the clouds parted, the rain stopped, and the "sun" came through the clouds—that is, it *flew* through the clouds, being not really a fiery ball too brilliant to look at, but rather a flattened disc shining like a pearl. It maneuvered around in the sky for a few minutes while all of the witnesses looked on—including some who were as far away as fifteen miles, having been unable to make the pilgrimage to Fatima itself. Finally, it became a brilliant, blood-red color, . . . It began to move rapidly toward the crowd below, appearing much larger and more fiery at each instant. The crowd, dried by the heat given off from the object, fell on their knees and cried for mercy, convinced that this was the end of the world. At the last moment, the Fatima sun halted its terrifying descent and retreated back through the clouds from whence it came. It was never seen again.[18]

Jacques Vallee explains that in the initial conversations with the Madonna, she gave enlightening information and instruction to the children. "It is also remarkable that the children were shown a vision of hell that terrified them and were given a specific prophecy announcing more appar-

18 *The Utah UFO Display,* pp. 198-99.

itions of unknown lights in the sky." The little blue lady (Blessed Virgin Mary) made further prophecies concerning the coming major wars and calamities of the twentieth century. This was as much a series of religious apparitions as any Catholic Saint had ever produced. Vallee is not convinced, however:

> The final "miracle" had come to the culmination of a precise series of apparitions combined with contacts and messages that place it very clearly, in my opinion, in the perspective of UFO phenomena. Not only was a flying disk or globe consistently involved, but its motion, its falling-leaf trajectory, its light effects, the thunder claps, the buzzing sounds, the strange fragrance, the fall of "angel hair" that dissolves upon reaching the ground, the heat wave associated with the close approach of the disk—all of these are frequent parameters of UFO sightings everywhere. And so are the paralysis, the amnesia, the conversions, and the healings.[19]

Hundreds of similar experiences have been reported from around the world. Either departed Catholic Saints have nothing better to do than put on sophisticated light shows for the faithful, or UFO occupants are counterfeiting Saintly apparitions. If this was a Saintly manifestation, what was the children's role in it and why could only they see the Madonna, while only the two girls could hear her? Another obvious question is: was this a spiritual or extraterrestrial event? As with the major contactee encounters of today, it appears that the encounters of yesterday were extraterrestrial with religious overtones or vice versa.

Blessed UFOs

Ed Conroy comments on the religious nature of UFO apparitions:

> An alternate hypothesis about the apparent connection between UFO phenomena and apparitions of the Blessed Virgin Mary holds that the majority of such cases are examples of the manipulation of mass human consciousness by entities capable of producing visual and auditory displays that are specifically designed to lead worshipers of the Virgin to believe that the displays are, in fact, the Virgin herself. The motivation for this deception, according to Salvador Freixedo, a Spanish ex-Jesuit priest, who does not blush at proposing hypotheses that would give the tabloid press a field day, is the entities' desire to psychically feed upon the mass emotions of agony and ecstasy that are only observed in miraculous shrines. Freixedo goes so far as to hypothesize that we are literally preyed upon by certain classes of entities who create UFO phenomena in

19 *Dimensions,* p. 200.

> the interest of feeding off of our religious emotions, in a manner analogous to the way in which blood has been found entirely removed from cattle that had been strangely mutilated in UFO related events throughout the world.[20]

Agreeing, Vallee attempts to focus in on what the phenomenon is: "The system I am speaking of, a system with mastery of space and time dimensions, will be able to locate itself in outer space. Nonetheless, its manifestations cannot be spacecraft in the ordinary nuts-and-bolts sense. The UFOs are physical manifestations that simply cannot be understood apart from their psychic and symbolic reality. What we see here is not an alien invasion. It is a spiritual system that acts on humans and uses humans."[21] UFO researcher John A. Keel likewise agrees that these paranormal apparitions, what he terms "soft objects," or "sightings of transparent or translucent objects seemingly capable of altering their size and shape dramatically," are "temporary manipulations of matter and energy."

But what are they? No one seems capable (or willing) of proposing a tangible theory. The only thing that these UFO researchers appear to agree on is that UFOs are *not* spaceships.

Ouija Boards And Seance Rappings

One of the most bizarre UFO contacts on record was experienced in eastern Utah. Again, we have Dr. Salisbury to thank for publishing the account.

> Sighting #41 is well calculated to frustrate the careful objective approach of a scientific UFO investigator. The Clyde McDonald family was informed by one of its members, a fourth grader, that a flying saucer was going to appear above the Roosevelt Hospital at eight o'clock on the evening of February 23, 1967. How did the young lady know that the flying saucer was going to appear? Why, some of the students in the other fourth grade class had been playing with a ouija board, and the ouija board had informed them that this event would take place. Mother and father chuckled properly and sat down to watch television. They were relaxing in their bare feet when the children began to bundle up and go out to meet their eight o'clock appointment with the UFO. Mother said, "You'd better hurry. It will be gone before you get there." Then, as Mrs. McDonald says: "Tim looked up, and he hollered, 'Oh, run, Tammy, There it

---

20 *Report on Communion,* p. 337.

21 *Dimensions,* pp. 284-85.

> is!', and they about broke that door off getting in. So we all ran, the three of us, jumped up and ran to the door, and, of course, then we could see that big light right out there by the Jennings' yard. Boy, they were behind us and they wouldn't go out then. And we saw that light, and, of course, we just stood out there then and watched it, and we about froze,—we were barefoot and everything, but we didn't want to come back in and lose it! It was kind of an orange ball, orange to red, kind of in a circle, bigger than an ordinary light." How big was it compared to a full moon?" I asked her. "Well, it wasn't that big. I'd say probably a third, wouldn't it be?"[22]

At least five other witnesses saw this same UFO from another location at the same time in the same place. Others claimed to have seen the object at slightly different times, moving, rather than stationary. Another witness claims that the object flew alongside his car for about a mile, then flashed vertically until it was out of sight.

What do we make of a UFO that announces an appearance at a certain time and place via a ouija board? When Dr. Salisbury first reported the case, his scientific approach seemed somewhat impotent to him.

Whitley Strieber speaks of nine distinct taps on the upper side of his cabin, in three groups of three, followed by two light taps. He ascribes much importance to the incident because he knows that the rappings were made by the visitors while he was in full consciousness. Strieber devotes an entire chapter to the rappings, convincing the reader that no person could have slipped by his security system to reach that portion of the house, describing his cats' unprecedented reaction, and depicting the rappings as so precise that there was a hidden message in them for him. The point that Strieber fails to appreciate is that such rappings and related phenomena have long characterized paranormal contact. There is no meaning or message—there is only rapping.

Steve and Dawn Hess report that after their initial encounter in the Mojave Desert, their family was tormented by poltergeist-like activities. Dawn and the baby experienced a rushing of cold wind through a closed living room and an invisible hand racing across the venetian blinds. Then one night Steve and Dawn rushed into their young son's bedroom to find him spinning in mid-air. Dawn grabbed him from the air and held him tightly.

---

22 *The Utah UFO Display,* p. 38.

Dr. Mack describes the following incident from the early life of "Sara." "In addition to the ghost story sessions, Sara used to play what she called 'seance games' with her childhood friends. Once at a slumber party, she asked her best friend, Annie, who was also the smallest, to lie down on the floor and said, 'We're going to try to levitate you.' I don't know where I knew about levitation either, and we went all in a circle. I think I was at her head, and I started saying something, and then it was like, now okay, and the girl went up, you know?' Each of the children who were present had 'a sense that something weird had happened,' and afterwards no one spoke of the incident. 'I remember that night very vividly,' Sara recalled. 'Oh, God! That night the whole room was very strange . . . There was a lot of electricity in that room. I think after that it was not even conscious for the kids.' I asked her if they told anyone about it. 'I don't think they even thought of the idea of telling.' It seemed to Sara as if there were 'a suggestion they don't tell.' A couple of years ago, Sara asked the girl she had floated, 'Did we lift you?' and the girl said yes and that everyone present was frightened by the experience." (*Abduction,* p. 204.)

Paul, another of Dr. Mack's abductees, related that he had inadvertently involved his therapist in the phenomenon. "As the sessions progressed, Paul found that he was becoming increasingly 'awakened' to 'a connection to like a foreign alien kind of thing' which Ms. T. [his therapist], perhaps understandably, could not deal with. One time in a session he asked for a sign of the reality of the beings, or the energies associated with them, which was followed by a loud bang near the office door. Ms. T. was frightened but willing to explore the noise. Paul was curious to check it out and felt an electric 'crackle' in the room but found there was nothing visible behind the door. Ms. T. was wide-eyed with fear and Paul had to try to calm her down. He had the sense that something was 'going to come' for Ms. T. over the next weekend and told her so. She volunteered no information at the beginning of the following session, so Paul asked her if anything had happened. Ms. T. reported that her bed had mysteriously bounced up and down. She revealed that she had been terrified, and, according to Paul, tried to ignore what had occurred except to clean the house 'out of evil spirits.' Her assumption, Paul said, was that anything 'good or intelligent would greet you in a very comfortable way.'" (*Ibid.*, p. 218.)

Photographing The Unknown

Joanne Wilson and her business companion, as reported in the book *UFO: End-Time Delusion,*[23] were followed by a UFO early in the morning as they drove to a swap meet. When asked if the UFO had a definable form she responded as follows: "It was like a cloud of gas or fog or something like that. Very dense though. But the light did not shine out. It didn't light the trees or the ground or anything. It was all contained within itself . . . . And there were big columns of light. They were white, white lights that came on inside it. They were so far from the bottom and so far from the top. They didn't touch the top or the bottom."[24]

Joanne and her friend stopped the truck, locked the doors, rolled down the window and attempted to take photographs of the UFO: "She took a couple of pictures there, and that's the two that turned out. And when we took the pictures, it dissolved, almost completely. It just dissolved itself. I thought to myself, *Whatever it is doesn't want its picture made.*"[25]

The photographs came out blurry, with white streaks instead of the clearly visible forms and columns seen distinctly by the two women.

This result is quite common in attempts to photograph other paranormal phenomena such as poltergeist manifestations. The photographer will see a clearly outlined manifestation, but the photograph will reflect only streaks of light. Additionally, the description of a brightly lit craft that fails to shed its light on surrounding objects is not only common in UFO reports, but also in other paranormal encounters. In the Gulf Breeze case, photographic analysis of the "UFO just above the road" photograph reveals that the light source was very bright at the point of departure—the UFO—but experts were puzzled that the light appeared to be confined to that specific area, and did not light up more of the road and surroundings. This is not to say that it failed to light the surroundings up at all, as in the case of superimposition of an image onto the film—but the radiated light is less than one would normally expect to view in the photograph. Personal friends of the author have recounted spiritual experiences wherein a visita-

---

23 David Allen Lewis and Robert Shrechise, *UFO: End-Time Delusion,* New Leaf Press, Arkansas (1991).

24 *UFO,* p. 23.

25 *UFO,* p. 33.

tion would be precursed by a sphere of light in a dark room. As one might predict, the light is contained, and does not radiate to illuminate the darkened surroundings.

New Age guru Ruth Montgomery quotes a New Age channeler in contact with an alien who explains such phenomena, saying that "the photograph was really an illusion; that it was not a photograph of physical matter, and humans were wrong in perceiving it as physical matter."[26]

Interestingly, prolific UFO photographer Eduard Meier relates that the aliens instructed him to just point and shoot his cameras, that focusing is unnecessary. Similarly, some abductees relate that they clearly view every detail of the abduction without the benefit of glasses or contact lenses. (See e.g., *Abduction,* p. 154.)

Ghostly Apparitions

The ability of the Gulf Breeze UFOs, as well as many others, to suddenly "wink out," or completely disappear, fits into this paranormal occurrence parameter. Other UFOs similarly appear to oscillate between this plane and another—seeming at first solid, then transparent. Betty Luca's opulating and deopulating UFOs appeared to be more than interplanetary industrial products. Is the ability to shrink a solid craft and its alien and human passengers superior science, or paranormal projecting activity? Why shrink such a craft and its occupants?—fuel economy? Hardly.

Dr. Mack shares Eva's accounts of strange encounters with spiritlike beings, wherein "she woke during the night and saw a 'violet rectangle, like a doorway into/out of somewhere not visible, maybe another dimension.' She saw then 'the upper parts of some people wearing white and they were standing at the front of my bed.' She blinked her eyes, thinking 'it's my imagination.' But 'when I opened them again they were still there. At some point I understood it's real . . . I felt as if they were 'bringing me back home'—whatever that means" (*Abduction,* p. 245). Soon, "Eva shifted from the direct reporting of her experiences to speaking of the motives of the aliens and the meaning of the abduction phenomenon, based on information they have given to her. 'Their purpose is to live in unison,' she said, 'not to take anything from us. They want to study us to see how they can

---

26 *Strangers Among Us,* p. 144.

communicate . . . There are different dimensions, worlds existing within worlds,' she added, 'and to go from one to the next is like a roller coaster. You need to speed up the energy, and then you go to another dimension where the reality is different. In the transition from one reality to another, you feel like you're contracting and expanding at the same time . . . It's like you become, on the one hand, part of everything, and everything becomes part of you,' but 'at the same time you contract into an infinitesimal point.' This, she said, is 'absurd, because it's two conflicting ideas,' but this absurdity contains the 'secret of moving from one dimension to the next'" (*Ibid.*, p. 250).

In the chapters that follow, the famous Men In Black phenomenon as it relates to paranormal activity will be discussed. It suffices to say here that the MIB are quite ephemeral in nature, almost supernatural, and that they are every bit as anomalous as any aspect of UFOs and their occupants.

Past Life Recall

Dr. John Mack's work has been the most forthcoming with indicators that the "aliens" are providing abductees with "past life recall" experiences. He reports:

> Past lives are experienced during the sessions with strong emotion appropriate to what is being remembered. This is most likely to occur when the investigator picks up on cues in the sessions during which encounters from infancy are being remembered. Complaints or simply observations of being here on Earth "again," of being "back" or having "returned," are voiced (about which I then inquire). The past lives that are recalled seem to have relevance to the personal development or evolution of the experience, as I have seen in the cases of Dave and Joe.
>
> Past life experiences provide abductees (and the investigator) with a different perspective about time and the nature of human identity. Cycles of birth and death over long stretches of time can thus be relived, providing a different, less ego derived sense of the continuity of life and the smallness of an individual lifetime from a cosmic perspective [*Abduction,* pp. 48-49].

Dr. Mack appears to be a believer in the reincarnation tenet of the Eastern/New Age religions. With abductees like Catherine, whose account follows, he feels that there exists strong evidence of the reality of reincarnation.

> Then [the aliens] showed her a current picture of tomb paintings with the paint flaking off, "but then it switched to me painting it." But in that incarnation she was a man and

> as she watched this scene "this makes sense to me . . . This is not a trick. This is like useful information. This is not them pulling a bunch of shit like everything else." Catherine now felt that her insistence upon a more reciprocal exchange of information had been affirmed.
>
> I then asked Catherine to tell me more about this image of herself as a painter in the tomb of an Egyptian pyramid. In response to my questions she provided a great deal of information that she seemed to know as the painter, who was called something like "Akremenon." Some of this Catherine could have known from reading standard books about Egypt. Other details, such as the process of making the paint which she seemed to know well, and appears to be consistent with one textbook account, would not most likely have been known to her. She described the man's skin color, his clothing ("just a wraparound loincloth") and headdress, which denoted the higher status of royal service when compared, for example, to a slave. What was striking at this juncture was the fact that the quality of Catherine's experience was totally transpersonal, i.e., she was not having a fantasy *about* the painter. Instead she *was* Akremenon and could "see things from totally his point of view instead of from me watching it." [*Ibid.*, p. 172]

Catherine was given a great deal of information during this reincarnation session with the aliens. She was absolutely convinced that she had lived as this Egyptian painter because she felt his thoughts and feelings.

Joe, one of Dr. Mack's subjects, was given multiple reincarnation episodes by the aliens. Upon dying in one life, he was then born into the next, reliving all of the details of his death and rebirth. "With intense feeling and conviction Joe told of being a poet named Paul Desmonte in a village near London at the time of the industrial revolution. Desmonte was arrested, tortured, and he died in prison after blaspheming against the political and religious establishment. I took Joe through the details of his arrest, prison experience, and confrontations with the authorities. In prison he was starved, kicked, and beaten with sticks and belts, which left him with broken fingers and ribs. . . . Then 'these ETs' came back. Joe attributed their return to his struggle to open to a greater sense of himself, to his letting go 'of my bitterness' and of his head-to-head battle with the authorities." (*Ibid.*, pp. 193-94). The "ETs" then helped Joe through his death as Desmonte, always there, guiding him on his New Age path.

Not stopping at convincing Joe that he had been other people throughout history, the aliens now set out to persuade him that he was also an alien in past lives, and sometimes, in this life.

> "Etherial," "fluid," and a sense of "vastness" were other ways Joe described what it felt like to be in the alien form. He felt "incredulous," doubting his experience and wondering how he could hide from himself that "I also exist here, on the ship . . . I'm just so much more comfortable." He then described an intense struggle between his

"humanness" and the humanoid identity, which he had maintained as separate. Yet he also felt that he was "more fully integrated in being human" than most alien beings.

Joe called the race of beings to which he belonged in his humanoid identity the "brotherhood" or the "Obasai" people.

Joe told then of an experience he had had just a few days earlier as "Orion," his humanoid identity. He felt himself to be seven and a half to eight feet tall, although he sensed that he could make his body taller or shorter. A blond woman of about thirty-five he called "Adriana" was brought to him so that he could "make love with her" and "give her my seed." Although Joe felt that "she's been involved a long time," it troubled him, at least in his human self, that a part of her was frightened. Adriana, Joe said, was walking her dog at night when she was abducted and was "in a sleepy state" when the beings floated her into the ship. "Part of her freaked out when she first saw the ship," Joe said. . . .

Adriana was placed on a slightly tipped platform with her head higher than her feet. She was kept in a sleepy or dreamy state ("mentally they create, like, this web . . . they just drape her in this soft, gentle energy") as she was undressed by small beings. "There is that fearful part of her that just totally doesn't want any of this to be happening," and when this resistance "bubbles up to the surface" the beings controlled her through a kind of energy "massaging." [*Ibid.*, pp. 184-85].

Like Joe, many of Dr. Mack's subjects are being convinced by the aliens that they are "double agents," half human and half alien, and that they carry out assignments for the aliens on a regular basis. This scene experienced by Joe is an indication of just how far these beings are willing to go to brainwash humans into believing that they are inseparably connected with the aliens and their antihuman assignment. Joe's case is not unique. "Sara" also received this level of convincing.

I asked Sara what made her feel that Mengus [her alien mentor] was dead. She replied, "I can just read his vibration, and when I go to find it now it's just like it's died and been recycled (see Paul's explanation of what happens when a being dies, chapter 10). I can't access him anymore and he feels dead." Mengus "was really nice. I would say, maybe like my first real teacher." She has "the weirdest feeling that one of the little things I drew, the baby ones, was . . . Mengus's new incarnation." (*Ibid.*, p. 208).

In this life her body was skeleton-like, "like Mengus's . . . It's creepy, and your bones are kind of little and brittle and it's kind of creaky. You walk in a very disjointed sort of way." Again, Sara was struck by the joy of the maneuverability she felt within the space vehicle, how "just neat to zip around" it was.

From this alien/past life perspective Sara spoke of the "stupid" things that humans do and the temptation to confront them [humans] directly. But "it's much more useful to be subtle and make sure they thought of it themselves." Human beings are "so egocentric they won't change otherwise. They haven't. They've got this ego thing that they like to hold on to and they get really threatened . . ." At the same time there are "precious" things about human beings. "They can smell flowers, for instance. And that's so incredible, and they get to feel the sun on their skin." As an alien being "I was operating out of less physicality, so you're lighter at one level . . . ." At the same time the aliens have seen "a bigger picture," and have more insight and patience. Also, "You [as an alien] have this thing in your head that [enables you] to access any kind of information telepathically. So you have this kind of informational pliability. I mean, you can get any information you need." [*Ibid.*, p. 209].

Eva also was given past life memories, which she experienced with great feeling.

In the other entry she described a past life as a teenager in the 1930s or during World War II. She saw a baby in her arms, and believed that this experience explains her love for all children. . . . At the beginning of the session Eva described how before a lecture at her energy and healing school she went to take a nap, could not fall asleep, felt restless, listened to music, "and then it started." She saw herself as a five- or six-year-old boy in the mountains somewhere in Europe. She lived with her father, a husky, blond-haired man, in a log cabin. She described their clothing, including the detailed designs on their white aprons. "We had something on our head like a yarmulke, like a cap or something." Turning to the left she saw a "saucer, spaceship." A few minutes seemed to pass, and "I was walking towards this spaceship and my father then and there was frozen. I mean, he couldn't move; he couldn't speak. He was like, you know, ice." As the boy, she saw "one of those midgets again coming out." Next she was in the ship, which took off as she looked down through the window at her father, who "unfroze" and looked up with tears in his eyes. . . . Somehow she knew the year was 1652 [*Ibid.*, p. 257].

She remarked about past lives in World War I and II and "in Morocco long before that." Her "drive" in each instance was "to help mankind overcome blindness." I asked about Morocco. She had been a rich merchant named Omrishi in the early thirteenth century who was trying to "undermine" corrupt officials of the hereditary ruling family that dominated the local government of the village. . . .

Men on horseback dressed in black with white headdresses—"the bodyguards of the ruling family"—came to Omrishi's tent to arrest him. The women around him cried and the children hid, for they knew what was going to happen. He was taken to a white stone building, which smelled foul from people puking and urinating there. Omrishi was to be beheaded, and the people were told to gather in the center of the village to watch the execution, for they wanted "to deepen the fear in them." After his arrest Omrishi was taken from his cell to see where the beheading would take place. The following morning at ten o'clock "they took me, they got my head on the thing, and plop." The feeling was one of "release, freedom," of "going up, expanding, joy . . . There is no description," she said. "All I feel is white light, gold light" [*Ibid.*, p. 259].

Dave also was made to believe that he had lived many lives, always under the guiding hand of the aliens.

> The first thing that came to his mind after a long pause was that he was a Native American boy, given the name Panther-by-the-Creek, of the Susquehannock tribe living near Pemsit Mountain, and he was studying to be a medicine man. This was in a time "before the Indians knew about the White man," Dave said later. The boy lived by the river, caught shad, and dried meat for the winter. Eagles lived along the river in the cliffs. "The eagle is real special. The mountain's like a special place. Medicine men go up there to get visions, do journeys." When he was ready, he too went up the mountain to get visions, and it was there that he met Velia (the familiar alien female), who was a "friend and protector." He felt sad, for he missed being with her and also because something was to cut his life short. He cried a little as he remembered the territorial wars with the Iroquois and "a big fight around Hollowman's Island." He wore deerskin and carried a bow and arrow and war club and was just starting out as a warrior.
>
> The battle was confusing, with a lot of screaming, and he was shot in the left side of the chest through the heart with an arrow. "It burned, and then it numbed. It was just numb." He coughed up blood, which filled his mouth and he choked on it. Then he blacked out and died. "The next thing I knew I was away from my body." He saw his body lying on its back below and also one of the Iroquois warriors bending over him and cutting off his scalp. Then he felt himself "floating up in the air" and "dissipating, spreading out all over like a fog of crystals . . . I just went everywhere. . . ." [*Ibid.*, pp. 284-85].

Dr. Mack says, "I asked Dave what his connection was with Velia after he died. He answered, 'When I floated out of my body, before I dissipated, she told me that she'd be with me even when I came back.' Again feeling sadness he said, 'Then I was born again in Virginia.'" Julia, another abductee who had been sitting in on Dave's hypnosis sessions, had been given past life memories of Dave's Virginia episode, and the two believed that they had lived as brother and sister in that life. The odds of this being possible, even accepting reincarnation as plausible, are quite low. Apparently understanding this obstacle to the truth of their revelation, like Rupert Sheldrake, Julia too speculated that more than one person "could potentially access the same life" (*Ibid.*, p. 288).

With this one point we can agree these past life recollections are not true abductee memories but are detached pools of data, like panoramic programming, prepared for the purpose of convincing abductees, gurus, and us that "God's Plan" is not the way things are, and that we must look to the "aliens" for spiritual progress and eventual salvation.

Panoply Of The Paranormal

Contactees and others hear telepathic voices and see vivid pictures within their minds; light beams are used to lift, immobilize, and transport; those who encounter UFOs have their will taken over by an external force; UFOs disappear and reappear as though outfitted with Star Trek cloaking devices; UFOs dash through our skies at tens of thousands of miles per hour, causing no wind or sonic booms; UFOs can see, hear, and control remotely; people who encounter UFOs are compelled to act; contactees are taken to strange new worlds, shown wonderful architecture and technology, and are floated over valleys and through walls; UFO occupants deliver religious and prophetic messages that are untrue; and, UFOs are able to change their size, shape, and appearance at will, as are their occupants.

If these same characteristics common to UFO encounters were referenced in a mystical or spiritual context—say, we merely eliminated the spacecraft itself from the equation, there would be no doubt that they properly fit into the paranormal classification. Their "presentation" in a superior science context fails to remove them from the paranormal realm. Furthermore, the historical presence of the phenomena only supports their paranormal underpinnings. In short, whatever UFOs are, they have always been with us and they exist outside of our everyday space-time continuum—in many cases, anyway.

Dr. Mack specifically noted the many paranormal aspects of the UFO/alien encounter. "These include the apparent mastery of thought travel by the aliens and sometimes by the abductees themselves (as Paul, described in chapter 10); abductees' sense that their experiences are not occurring in our space/time universe, or that space and time have 'collapsed'; a consciousness abductees experience of vast other realities beyond the screen of this one, beyond the 'veil' (a word they frequently use); the deeply felt sense of opening up to or returning to the source of being and creation or cosmic consciousness, experienced by abductees as an inexpressibly divine light or 'Home' (another word they commonly use); the experience by abductees of a dual human/alien identity, i.e., that they are themselves of alien origins (for example, Peter, Joe, and Paul, in chapters 13, 8, and 10); and the powerful reliving of past life experiences, including great cycles of birth and death. In addition, the aliens appear to be consummate shape-shifters, often appearing initially to the abductees as animals—owls, eagles, raccoons, and deer are among the creatures the abductees have seen initially—while the ships themselves may be disguised as helicopters or, as in the case of one of my clients, as a too-tall kangaroo that appeared in a park when the abductee was seven" (*Abduction,* p. 32).

Certainly, whatever their source, the UFO phenomenon and traditional paranormal phenomena appear to have common origins. Herein we have only scratched the surface—any researcher, or even curious observer, who reviews the prolific alien encounter reports that inundate the media must accept the fact that such encounters, when reduced to their essential (ghostly, psychic, spiritual, religious) encounters without deviation.

The question is: are UFO and alien encounters a new branch of traditional paranormal activity, or vice versa? In the remaining chapters we explore the stark probabilities.

12 Who Is Sending The Message?

Isn't It Obvious?

The overriding question behind the UFO phenomenon is who is sending them, complete with a message to mankind? The apparent response is that they are piloted by humans or humanoids from another planet. Yet, as we examine the nature of the UFOs themselves, as well as the "message" that they appear to be sending, the apparent answer becomes less evident.

A Paradigm For A Plethora

In this chapter we reexamine the possible sources of UFOs, weighing the supporting evidence for each such source in the balance of truth. UFO debunkers believe in nothing that cannot be fully documented on the network evening news, especially extraterrestrial life or spiritual entities. They, therefore, accept only natural phenomena (meteorological, electromagnetic, psychological) as possible explanations of UFO sightings and encounters. New Agers believe that UFOs are piloted by, or are manifestations of, more highly advanced spirits that have come to Earth to usher in its next step in human evolution—the Aquarian quantum leap. Other observers have come to believe that UFOs and their occupants are extradimensional—hailing from parallel universes or other time periods in our own universe, fading in and out at will. However, most UFO researchers and "fans" fall into the "extraterrestrial explorer" category, accepting at face value that we are indeed being visited by beings from other planets. They appear to accept the practical aspects of the UFO message—that Earth is becoming polluted and will require alien help to overcome an impending cataclysm, and that they are performing genetic research to preserve our race and/or theirs.

Theory 1: Extraterrestrial Explorers

The extraterrestrial explanation is the most widely accepted in our society. Indeed, UFO occupants traditionally declare themselves to be humanoid inhabitants of other planets, arriving in nuts-and-bolts flying machines built with engineering capabilities acquired over tens of thousands, or even millions of years of scientific inquiry. In support of this hypothesis we have millions of independent witnesses telling believable stories of sighting solid, reflective craft that maneuver at incredible speeds, performing feats that are deemed impossible by the standards of our current level of technological understanding.

Hard Evidence

In addition to the numerous sightings, we have evidence of the physical presence of UFOs. This evidence, however, when considered in tandem with other reported properties of UFOs, leaves experts unsure about their nature. Scientific inquiry is almost impossible under present circumstances, and mere analysis of the available data sheds little light on the problem, as observed by James McCampbell as recorded in the *SCP Journal.*

> Physicist James McCampbell, speaking to a UFO symposium in 1975, rendered a succinct analysis of the problems encountered in a scientific study of UFO characteristics: "Evidence left at landing sites leaves little room for doubt that UFOs are heavy, ponderous objects when at rest. Yet in flight, their startling departures, sudden stops, and right angle turns at high speed require them to be virtually massless."[1]

This seeming contradiction is explained away by "extraterrestrial theory" adherents as merely superior science. The fact exists that there have been many well-documented landing sites throughout the world. Although insufficient to convince debunkers of the reality of UFOs, the physical evidence of their presence is mounting. In addition to these material signs of the physical reality of UFOs, we have numerous photographs. All of this, however, can be counterfeited at one level or another.

Photographs

Ed Walters's photographs and videos are among the most convincing that UFOs are actual, solid, real objects. There exist many others as well. In

---

1 *SCP Journal.* August 1977. Vol. 1 No. 2. "UFOs—Is Science Fiction Coming True?", p. 14.

addition to those mentioned above who have taken movies and videos of UFOs, television News and "Mystery" program film crews have photographed what they believe to be UFOs on high quality motion picture film as well as commercial-grade video tape. The weekly television program *Sightings,* which covers subjects in the realm of the paranormal, documented a research crew's efforts to film UFOs in one of the latest UFO "hotspots," Anza, California. Citizens of that town report an escalating number of close encounters, including abductions and breeding experiments. The film crew set up its cameras in a place well known for sightings, and within two days was able to film a UFO moving through the sky. Of course, no attendant documentation or investigation was available at the time of airing the program, so it was only assumed that no other explanations existed for the presence of the strange flying craft. The effort does demonstrate, however, that UFOs are becoming bolder with each passing year. Similar efforts reaped comparable results in Gulf Breeze, Florida. Television programs receive enough videos depicting UFOs that their audiences are treated to weekly displays accompanied by experts analyzing the videos.

Prime Time Videos

One late night "news" program (frankly, not known for its journalistic professionalism) aired in late May, 1992, a video sent back live from the space shuttle that was indeed intriguing. A UFO researcher was analyzing the video as it was replayed again and again. On the screen we saw a view of the Earth, toward the edge so that we saw the horizon to the right, and outer space beyond that. Travelling over the Earth from the bottom of the screen to the top within a few inches (on the television screen) of the right horizon was what appeared to be an unanimated object. NASA had just commented that the object looked much like an orbiting hunk of ice (very large), expelled from this or another mission. This particular object appeared to be orbiting at several times the speed of the rotation of the Earth, which is approximately 1,000 miles per hour—the author's layman's guess would be that the object was travelling at 40,000 to 50,000 m.p.h., possibly much faster. Just as this object was about two-thirds of the way up the screen, still travelling at the same high velocity, it made a perfect right angle turn toward the right of the screen, moving toward outer space.

This it did without any interruption or hesitation—it changed direction instantly at 90 degrees as though it had been a billiard ball rebounding off of a table cushion.

If this maneuver were not exhilarating enough, within one or two seconds of the object executing its amazing right angle turn, a second object shot up from the direction of the Earth's surface through the point of intersection where the first object would have been had it not made its seemingly "evasive" maneuver. The second object quickly shot off the screen toward the top, while the first continued on its course to the right of the screen into outer space, quickly fading from view. The second object was moving even more rapidly than the first, and because of its "streaking" appearance, it seemed that it could have been some kind of energy burst rather than a solid object.

The scene was extraordinary as it replayed over and over on television. The UFO analyst shared obvious observations about the speed, trajectory, and apparent tactical implications of the display as we watched it repeatedly. It was an amazing display, and does much to support the nuts-and-bolts UFO theory—after all, as far as we know, we have no such vehicles flying around out there that can maneuver in such an incomprehensible manner.

In this space shuttle spectacle, the second object is of almost greater interest than the first. If, as it appeared, the second object was "fired" at the first—who launched it and what was it? Did our government shoot something at an unknown object, so close to our own space shuttle? It does not seem likely. Neither does it seem plausible that one of the few foreign governments with sufficient technology to fire such a projectile would do so with the shuttle so near. If so, does this imply that governments of the Earth are in active combat with UFOs? If this is true, then we are being kept in the dark about a serious enemy and technology well beyond our current understanding. If not governments of the Earth, is someone else in active combat with UFOs? Who? Could it be "good" and "bad" UFOs fighting over the Earth? Could it be hidden Earth civilizations locked in combat to protect us all? Although this entire discussion will seem quite sophomoric to some, these scenarios are precisely appropriate if the space shuttle video accurately recorded what appears to be a high-velocity right-angle turn to avoid an incoming projectile.

Target Practice

Whether or not world governments are locked in combat with UFOs, another apparent piece of evidence that UFOs are solid objects comes from the projectile-launching deer hunters of the Uintah Basin of eastern Utah. At least two hunters report taking shots at UFOs with their rifles and hearing the ricochet as the bullets glanced off the seemingly metallic objects. In both cases the UFOs left immediately, and in at least one case, returned and followed at a distance.

Personal Testimony

In addition to the millions of sighting witnesses and mounting physical evidence of the tangible presence of UFOs, we have the testimonies of thousands, including civic, political, and church leaders, who claim direct, personal contact with UFO occupants. Not only do these contactees and abductees convincingly relate the spiritual or emotional feelings (horror or overwhelming love) that they felt during their ordeal, but they are able to describe with great clarity the features of the beings they encounter and the nature and properties of the craft, and their means of operation. These abductees are absolutely convinced of the reality of their encounters with strange beings in foreign craft, and relate many physical aspects of their experiences.

The Personal Touch

For example, many claim to be physically touched, moved about manually, or examined or operated upon by their abductors. The material aspects of such personal encounters tend to place them in the real, physical realm. Udo Wartena recounts how he shook hands with those whom he encountered who professed to be from another planet. Others relate how they feel the cold touch of the beings' hands as well as the pain of the chilled, probing instruments, which leave cuts, scrapes and scars the morning after.

If we can use Udo Wartena's encounter as a standard by which to measure close encounters generally, there appears to be a clear division of encounter types. The first is the kind in which the contactee is fully conscious during the entire encounter (although Udo reports becoming unconscious upon the craft's departure) and reports many physical aspects about the encounter. The second would be the semiconscious bedroom visitor type. The distinction becomes muddled, however, because percipients of both groups often are unconvinced of the physical reality of their respective experiences. They sometimes "know" that it could not have occurred in physical reality.

From Udo Wartena's encounter we learn that handsome and youthful appearing humans ranging in age from 600 to 900+ years travelled here from other planets in a saucer-shaped craft supported on retractable legs. The age of the human visitors and shape of the craft are similar to the reports of others. As also related by others, Udo's visitors were here to gather water through hoses for their craft and to do research on cultural progress and pollution levels of the Earth. Udo likewise felt overwhelming love in the presence of his visitors, and was told of their "noninterference" policy, as is related by many other contactees. Finally, Udo's extraordinary description of the craft and its technology and means of propulsion is very similar to many other accounts. His explanation of the craft focusing on the energy of a distant star and skipping upon the light waves at ultralight-speeds was given decades before most people had ever heard of Einstein or his pertinent theories.

There are a few ways in which Udo's experience differs from those of others. For instance, although some report hearing loud noises at landing and takeoff as did Udo, the majority relate a stillness that falls over everything, with the possible exception of a "hum." Again, although many report encountering completely human occupants of UFOs (generally benign), the majority identify the occupants as the "small grey" variety, or some other variation of the gnome or humanoid forms. Udo specifically queried his visitors concerning their knowledge of Jesus Christ, to which they merely responded that they were not at liberty to discuss such matters pursuant to their noninterference directive, despite their desire to do so.

Unlike these, Betty Luca's visitors, as well as others', talk extensively about the spiritual or religious aspects of their missions—in seeming violation of their self-declared policy of noninterference. Although others have recounted cordiality between them and their visitors, as does Udo, most of those claiming to have such intimate contact with UFO occupants relate nonpermissive experiences with tranquilizing mind control and physical examinations or operations. No genetic samples such as blood, tissue, or semen specimens were removed from Udo. And finally, although Udo's visitors invited him to continue their journey with them, an offer not unprecedented in the literature, most receive no such proposition. Generally, "spokespersons" selected by the Space Brothers claim a UFO ride or two in our solar system, but rarely do we find such a bonafide offer of long distance journeying. Finally, and perhaps most significantly, Udo was not a "repeater"—he never saw his extraterrestrial friends again. If there is true "extraterrestrial" contact being made, we would culturally and socially expect it to be like Udo's experience.

Anything Is Possible

Even though we ignore some glaring red flags in our discussion of the possibility that UFOs are actually of extraterrestrial origin, there is nothing that necessarily precludes such a conclusion. Indeed, we can assume they are out there on their own planets, in great numbers. The only questions are, do they visit us? and if so, are they in the UFOs reportedly seen by millions worldwide? Is there some unknown logic or hidden meaning behind their apparently nonsensical message? Certainly, millions of trustworthy witnesses bear most solemn testimony that the craft and occupants are real, physical objects. We would like to think that all of the UFOs reported are piloted by benign human brothers from fellow earthly spheres, here to extend a warm hand of fellowship. This is *possible*—but not probable.

Theory 2: Returning Earthlings

There are those who speculate that UFOs are the flying machines of those who have (or will have) departed from this Earth, and who have returned for some unknown purpose. The dominant theories are: an ancient, advanced Earth civilization that once left the Earth is now sending back scouts: and, time travellers, from our past or future are here on a mission to observe us or to save the planet and mankind from certain self-destruction.

The "returning earthling" theory is that an ancient, advanced Earth civilization that once left the Earth is now sending back scouts for some unknown purpose—presumably, according to the message, to avert impending disaster here or to reintroduce themselves as our departed brothers. Many of the possibilities relative to this theory are weakened by the two questions: was there sufficient technology had at some time in the past to enable such an exodus? and if so, why do most UFO occupants appear significantly different from us?

Because of the numbers of sightings and encounters, the hostility, and the message, the possibility of UFOs belonging to an ancient, departed or future Earth civilization is remote. However, this does not absolutely preclude the possibility. There is no limitation that we know of on the existence of numerous, hostile encounters with departed Earth tribes that deliver nonsensical messages. Certainly, analogous circumstances exist among Earthbound peoples.

Theory 3: Natural Phenomena

The Debunkers' theory that all UFO sightings and encounters (as well as scriptural encounters with Deity and angels) are explainable by naturally occurring phenomena fail to address the hard physical evidence or the encounter/abduction phenomenon. Debunkers can blindly attempt to convince the world that millions have been mistaken in thinking that they saw silvery disks flying erratically in the sky, but this position is purely untenable at this point in the phenomenon. Too many trained observers have carefully watched UFOs perform their stunts from close range to be fooled by flocks of geese, meteors, or swamp gas. Mass hallucination is itself a pipe dream.

Debunkers proffer theories that Jungian archetypes common to all humans, or possibly birth trauma memories, are responsible for psychological ordeals like these that are similar, yet experienced by millions of persons with little else in common. The range of witnesses is far too wide to draw any cultural or psychological boundaries around those who see UFOs and abducting aliens and those who do not. Clearly, the Debunkers do not have the answers.

Because of the interrelated nature and complexity of the remaining theories—that the UFOs and their occupants are extradimensional or spiritually evolved entities—we discuss them together in the separate chapter, which follows. As for those theories treated in this chapter that attempt to explain the nature and origin of UFOs and their occupants, our only conclusion can be that they are *possible, but not probable.* This is not to say that some of the encounters cannot be fully explained by the extraterrestrial theory or some other theory discussed here—it merely indicates that none of these theories is inclusive enough to accommodate the wide range of behavior of the UFO phenomenon. This may indicate that different phenomena are being manifested—perhaps, a mixture of theories we are discussing is the answer. It is possible that many witnesses see robotic craft from other worlds in our skies, while others encounter true extraterrestrial explorers. Yet, we cannot believe that many of the encounters we discuss are answered by these explanations. Many "alien" encounters may be better explained by the theory we explore in the final chapter.

13
Are UFOs And The Alien Message A Satanic Deception?

In this chapter we discuss the possibility that UFOs and their related phenomena originate from spiritual sources—more specifically, whether they are of Satanic origins. This is the least flattering of all of the possible sources that are discussed herein, yet the most pregnant with opportunities for analysis. Christian Fundamentalist writers dismiss the entire UFO phenomenon as a demonic device *ab initio.* This is due not to their thorough analysis of the facts, or to claimed divine inspiration, but to their belief that Earth is God's only creation on which he has placed his finite number of spiritual/physical creations—man. With this understanding, Fundamentalists have no place in their theorizing for extraterrestrial humans; their reasoning is that the UFOs, if not built by Earth humans, must necessarily originate from the only other available source of intelligence in the universe—Satan.

Counterfeits Are Patterned After Truth

Although our general population is not as restricted in its theorizing as are Christian Fundamentalists, we must genuinely consider the real possibility that UFO phenomena are demonic in origin. The evidence of such is great. However, this is not to say that the Earth is not visited occasionally by our extraterrestrial brothers even if the bulk of UFO encounters turns out to have Satanic origins. Scripturally and traditionally we know that if Satan sets out to deceive, he does so by manipulating truth. If the majority of UFO sightings are a Satanic device employed for a purpose discussed at length later, or some other unknown purpose, Satan having selected it as a deceptive device supports the proposition that his manifestations are counterfeits of true occurrences.

Satan's *modus operandi* in his efforts to undermine God's plan has always been to confuse, divide, and distract God's children. His tactics are like those of the famed fictional courtroom attorney, Perry Mason. If a witness saw Mason's accused client at the murder scene, Mason did not directly attempt to convince the witness that he did not "know" what he saw. Instead, Mason employed an "Acme Models girl" to dress like the client and stage a second appearance before the witness to "implant" a new memory of the original witnessing. When the witness identified Mason's client as the one seen on both occasions, Mason put the counterfeit on the witness stand, who swears that it was she who encountered the witness on the second occasion. This technique, of course, confuses the witness and has the effect of rendering the witness's testimony invalid as to both events. This is "new reality" creation at its best.

The world's religious literature both ancient and modern, is filled with accounts of religiously symbolic personages (e.g. Virgin Mary) appearing to the faithful and presenting nonsensical messages from God. Satan sends these counterfeit messengers—glorious, radiant beings, praising God and commanding the percipient to worship God the Father and His Son Jesus Christ, and none other. The messenger of light then often proceeds to deliver a message that is 95% true. The five percent portion of the message, however, is generally deadly in terms of God's will in the matter. Examples are found worldwide, throughout the past two millennia as "Blessed Virgin Mary" apparitions. Satan sends these counterfeits for the same reason Perry Mason does—to confuse what we innately know to be the truth. They appear and act like the real thing, yet their message is distorted just enough to make the truth a lie, and enslave clamoring faithful to false traditions.

Power Of The Air

Satan is called the "the prince of the power of the air" in the New Testament by the Apostle Paul in (Ephesians 2:2). What does it imply to call Satan "the prince of the power of the air"? We are nowhere given the answer to this question, but in light of the escalating "signs and wonders" that are zipping through our skies, it would not be too difficult to form a hypothesis about Satan's abilities to create a convincing display in the heavens.

Deceptive Devices

We raise the devices of "theater," "counterfeits," and "*knowledge* distracting implants" above because as we analyze the UFO phenomenon from the demonic origin viewpoint, we should view each aspect of the UFO encounter in light of how it could be the implementation or employment of one of these devices. We are not the first to consider these issues. Many observers of the UFO phenomenon have remarked how closely related it appears to be to phenomena associated with demonic hauntings, apparitions, and bodily possessions. Even Dr. Salisbury, in the early 1970s, noted how the UFOs seemed to be putting on a "display," something that a scientist was hard-pressed to explain. Of course, a UFO's communication with the Uintah Basin children through an Ouija Board was an occurrence that he did not even know how to approach in a scientific medium.

Generated Images

Looking first to the physical appearance of the UFOs and their occupants can be enlightening. Much of the literature documents how UFOs initially "resolve" themselves from a dispersed mist into a sharply "focused" object. They then zip and zigzag about at phenomenal speeds, only to "dissolve," or disappear in a blink. As in the cases of the French Dr. "X" or Betty Luca, some UFOs are reported to transform their shape or size, or even merge with a second UFO as the witness looks on. The cry of "superior science" is drowned out by the roar of "projected image" when it comes to these kinds of displays. An image generated by a simple slide projector can travel quite quickly and erratically along a wall by the slight pivoting of the projector at its source. Can this rudimentary concept be applied to UFO sightings?

There are those who believe that UFOs may be nothing more than mere projections—laser or holographic. Although this theory is possible and interesting, we have little evidence that anyone possesses the level of technology required to put on such a light show. However, the technology is not too far in advance of state-of-the-art projection capabilities, and the possibility should not be overlooked. The question then arises: who would go to such extravagant, sophisticated, and expensive lengths to fool so many people? Only persons looking for power, as we discuss below, would conduct such a campaign. This would limit perpetrators to the political and/or religious realms.

Although this theory seems workable, it is not a practical one. The risk of being caught in the act of setting up the necessary equipment is too high for never having been discovered. This is especially true in light of the massive investigation that attends many UFO sightings. Besides these difficulties, this theory fails to account for victims' close encounters with, and examinations by "aliens," unless we include the use of drugs and hypnosis in the equation. Then, however, we have the radar and photographic images that have been produced by these phenomena to explain. UFOs seem to be solid objects—semisolid at least.

Spiritual Projections

If not laser or holographic projections, then what are UFOs? Does Satan or demonic beings generically have the ability to resolve a semisolid image before a human, complete with a messenger delivering a message? As so many witnesses (including Biblical prophets) have testified, the response is a resounding "Yes." How do they do it? We can only guess, but we know that they can. Dr. Curt Wagner, a physicist whose doctoral degree was earned in the field of general relativity theory, explains how he believes it is done:

> Drawing from what we know can happen in seances and poltergeist activity, it seems that these supernatural forces can manipulate matter and energy, extracting energy from the atmosphere, for example (which manifests as a local temperature change), to manipulate matter and produce an apparent violation of the second law [of thermodynamics], and I guess my feeling is that on a larger scale this is what a UFO could be. I'm not saying I know that it is, but only that it could be. It seems to me likely that UFOs are largescale violations of the second law in which energy is arranged to take on enough of a force field appearance so that it appears to look like matter, yet it's really just energy concentration—it's not really solid matter in the usual sense.[1]

Another researcher similarly concludes: "Demons, as fallen angels, apparently retain great powers, such as the manipulation and restructuring

---

1 *SCP Journal*, August 1977, p. 20.

of matter, as well as the ability to influence or control human consciousness and experience through classic possession or by direct psychic implantation of a set of experiences."[2] The extraction of "energy from the atmosphere, for example (which manifests as a local temperature change)," refers to the classic "cold spot" that is experienced during negative spiritual manifestations. As discussed below, UFO phenomena are often accompanied by such cold spots. Often, in fact, the larger the scene, the colder the temperature.

Catherine, an abductee being studied by Dr. Mack, noted that her abductors had an uncommonly "cold" feel to them. "Catherine became increasingly distressed, panting and crying, as she described how one of the beings spread her legs apart on the table and the examiner stared at her face and genitals. She noted that she had no clothes on. The examiner 'says something to the one on my right, and the one on the right goes off to the right side of the room to get something, and the examiner puts his hand on my leg, on my thigh, and it feels cold—not like human cold hands. I mean, it's even colder.'" (*Abduction,* p. 155.) Peter, too, reports the cold associated with his captors. "His attention shifted again to Nantucket. He sees a being floating toward Craig's bed and becomes anxious for him, but 'then it's time to go . . . I'm always cold when this happens,' Peter said, but also fog came in the room, and though 'it chills me,' it also 'calms me down'" (*Ibid.*, p. 317). This calming, cold fog is exactly like that reported by Steve and Dawn Hess in their Mojave Desert encounter.

New Age spokesperson and a prominent New Age writer Brad Steiger frankly admits: "I have even come to suspect that, in some instances, what we have been terming 'spaceships' may actually be a form of higher intelligence rather than vehicles transporting occupants . . . . I feel, too, that these intelligences have the ability to influence the human mind telepathically in order to project what appear to be three dimensional images to the witnesses of UFO activity."[3] Steiger's suspicious observations appear to support our theory. He, however, does not assume that the spirits involved are Satanic. His explanation, although tentative, is more innocuous: "I cannot help questioning whether the Space brothers might

---

2 *Ibid.,* August 1977, p. 19.

3 Brad Steiger, *The Fellowship,* Ivy Books, p. 49.

not be angels, spirit guides, and other messengers hiding themselves in more contemporary, and thereby more acceptable, personae."[4]

Certainly, history, through scriptures and personal accounts, tells us that spiritual beings, good or evil, have the ability to deliver messages packed with powerful images that fill and activate all of the human senses. For example, the Apostle Paul speaks of being taken up into the third heaven and being shown marvelous things, "whether in the body, or out of the body, I cannot tell" (2 Corinthians 12:1-4). Interestingly, in this same discussion he reveals that he had suffered the buffeting of "the messenger of Satan," for which he sought God's protection, which did not come easily. Paul also saw a light from heaven and heard a voice, but those with him heard no voice, although they saw the light (Acts 22:6-9). Peter, also an Apostle, had a vivid vision in which he saw "a great sheet knit at the four corners, and let down to earth: Wherein were all manner of fourfooted beasts of the earth," and he heard a voice (Acts 10:10-13). The Apostle John reports that during a full sensory panoramic vision with voices and conversation, that he "took the little book out of the angel's hand, and ate it up; and it was in my mouth sweet as honey: and as soon as I had eaten it, my belly was bitter." (Revelation 10:10.) The scriptures are filled with similar accounts of such full sensory psychic projections, originating from spirit beings.

From the accounts of abductees, we know for certain that the "alien" beings have this same ability. Steve and Dawn Hess shared the same "visions," testing each other's perceptions of what they were seeing and hearing during the eight hour ordeal, watching as thousands of small creatures swung through the trees and brush. Yet, the following morning they could find no trace of a fallen leaf or snapped twig; not even a single footprint. Steve Hess felt extreme hunger and suffered tremendous pain as his imagined hunters sent a bullet searing through his flesh in a psychic projection, as his wife looked on helplessly, watching Steve's limp body stare blankly in a trance. Scenes of humiliation, exultation, joy and fear were forced upon them, stolen from their own lives, as they relived every sight, sound, feeling, and detail of such scenes. Some of these scenes were reproduced with exact detail, and then altered for the purpose of provoking high emotions from the two.

---

4 Brad Steiger, *Revelation: The Divine Fire,* A Berkley Book, p. 148.

In an interview with the author, Dawn Hess described the great detail of her "altered state" bedroom abductions. Every aspect of the scene seemed completely real. She took great care to observe the minute characteristics of the furnishings, the lighting fixtures, and the inside of the UFO, or whatever the structure was to which she had been taken. She recalled the solid, smooth, cold feel of the metal chair and table. With labored effort she studied a styrofoam cup that one of the beings handed to her. It had the appearance and feel of a styrofoam cup that one would purchase at a grocery store. She noted the lip of the cup, rounded on top and squared along the bottom of the ridge. It was entirely indistinguishable from a normal coffee cup.

Similar full sensory scenes are reported by thousands of abductees, many of which are detailed in this work. Often, these projections are in the context of past life recall. In this regard, we must remember that millions, possibly more, of Eastern religious practitioners as well as aboriginal initiates (including Native Americans), receive these same psychic projections from some supernatural source. This source identifies itself variously as earthspirits, or advanced spirit guides, or aliens, or as messengers from God. At this point in our analysis, we should be prepared to entertain the probability that whoever these beings represent themselves to be, they are of the same source, not many sources, and are not sent from God. If these are demonic beings that are generating psychically transmitted scenes, which we know to be possible from other contexts, then they are expending a great deal of energy and resources. The magnitude and accelerated frequency of these transmissions tells us that there is an important purpose behind it all.

Physical Spirits

The question has often been raised, is Satan capable of physically intruding on our material world, manipulating objects with mass, and "touching" or otherwise physically affecting humans?

Those who have had the misfortune to experience demonic attacks bear witness to the fact that there exists a very real physicality to Satan's powers.

Anyone who has seen the movies *The Exorcist* or *Poltergeist,* or other motion pictures of that genre, have viewed synthesized versions of what is reported throughout the world. There is a very intrusive, physically abusive nature to demonic visitations. Ministers, Priests, Rabbis, and other church leaders, as well as anyone else who investigates such phenomena, bear solemn testimonials regarding the vicious violence associated with demonic attacks. The best witness is perhaps the Bible.

Such attacks are prolific, and much more wide-spread than thought by most people. A brief interview with one's own spiritual leader, or a cursory investigation among friends and family, will convince the reader of the escalating problems of demonic attacks.

Dr. Mack relates the following account of Paul. "Going in and out of sleep he felt sure that there was 'something in the room.' But when he felt 'something on top of me,' and found he could not move at all, he became terrified. 'I felt it further off, but then it came really close and, like, Whoa! That's too close. Too close! Back off! . . . As I'm freaking out, I'm trying to figure out where this thing is, and I was measuring it, like off the tip of my nose. . . . I could feel it right there . . . about four or five inches off my nose.' He wanted to open his eyes but simply could not do it. He 'dozed off again' (a common paradox in abduction-related terror) and then upon awakening was able to 'break out of it' and look, but there was 'nothin' there.' Although fully conscious now, Paul found he was 'gagging on my own breath because I couldn't speak' and was still unable to move for a few moments. Although he saw no beings during this incident, Paul nevertheless felt invaded by 'something [that] had definitely done something'" (*Abduction,* p. 221).

This was not Paul's first encounter of this kind.

> The scene changed and Paul went back to when he was a boy. He was about twelve years old in the cellar of his home when this episode began. "I'm fighting. I'm fighting alone. The thing that's fighting me knows that I'm right here, but I'm protected in some way, I guess. 'Cuz it can't just come out and just kill me. I think it wants to, though, definitely, he thought as it seemed to lunge toward him, only to be stopped by some sort of protective barrier. But it can't do that. It has to do it in other ways. It's going to try. It's going to take me apart, little piece by little piece . . . We've done this battle before. That's why I'm still here. He says he fights with everyone like this." The battle seemed mythic to Paul, as if he were "confronting destruction that wants creation stopped (some people call it Satan)," he said. "I'm yelling," he said, "but I don't think anyone's around."
>
> In the darkness Paul perceived a glowing, nonhuman creature staring at him that he called "'symbolistic' . . . It controls humans to a great degree" and wanted to "destroy" him. But he was protected from "getting lost" by a "creational [Godly] force" that held him. Once again he experienced numbness spreading through his whole body and helplessness. . . .

> No longer at home in his cellar, Paul now was lying on his back in his bed at night and experienced '"things' . . . moving around me." He could not move at all and saw "a thing in the closet" with a "horror face," like "that character from The Exorcist," that frightened him intensely as a child. The figure seemed to turn a light on, but receded further into the darkness of the closet. [*Ibid.*, pp. 231-32]

Are these descriptions of demonic spiritual attacks? Although this seems clear from the content, Paul is actually describing them to Dr. Mack in the UFO/alien abduction context. Even Paul perceives that Satanic and Godly forces are at work in these scenes of spiritual conflict, although his attackers insist they are associated with "aliens." Hearkening back to the Roper Organization survey, we found astoundingly high percentages of American adults have encountered these kinds of strange beings and phenomena. Also, the survey demographics reveal that the younger the adult population, the more frequent the paranormal occurrence. In other words, the frequency and intensity of these phenomena are escalating.

Many of Dr. Mack's abductees report physical attacks by invisible beings. Sheila, another abductee related the following. "Just three nights before at about three o'clock in the morning, Sheila had had another dream 'like the electricity' in what she called a 'semiconscious' state. 'I woke up real quick, and I could hear someone breathing funny and I felt a hand on my side.' She jumped up, screamed in terror, turned on the light and saw no one there. . . . [S]he recalled waking at about midnight and again at a quarter of one, hearing a noise, and feeling 'like someone had their hands on my arms and my legs. They stretched me face down on the couch, they stretched me out. I sat up and didn't see anything. I was scared out of my mind. I was terrified to be alone'" (*Ibid.*, pp. 84-85). This was not the only experience of this nature, Sheila complained.

Dr. Mack similarly relates the story of how one of these semi-solid "interdimensional" beings, representing himself to be connected with the "aliens," deceived Sara into giving up a little virtue.

> Sara associated to another experience later in her life. She was alone, lying down on the deck to get a suntan, when "I felt something hovering on top of me." She saw a figure that "was like a cross between a Mengus being and a person." It was "human in shape, but lighter and free-floating." Sara received a communication from the being, "This is very important." The intention, she was told, was not aggressive, but some sort of test of "genetic compatibility or something," an infiltration," "a feasibility test," "dimensional merging."

> I asked Sara to tell me more of what she meant by "dimensional merging." She then described what I believe to be the central image of our first session. "It is like a plane," she said, "a sheet of translucent cellophane." There is "like a huge shattering of glass," and a "razor blade thin" slit opens between this Earth/physical dimension and the realm from which the beings come. In this context I asked her more details of the encounter. The being had "a light contour of a penis, but not like a physical penis" that entered her body. The experience was not like anything she had known in human sexual relations. "The being itself felt aggressive, and I did not like that part of it. . . ." After this experience Sara "felt like I'd sort of been hoodwinked." The being "didn't give me the full story, and it just kind of said, 'Hey, trust me, it's important'" (*Ibid.*, pp. 210-11).

Scott, also a Dr. Mack abductee, learned that the line between alien abductors and demonic tricksters is too blurred to distinguish. "In that session he told me that one night about ten days before, as part of his increasing openness to the alien presence, he asked the beings to 'show me a sign' of their actual existence. At about two or three in the morning he experienced in a partially awake state the feeling of 'somebody touching me from behind.' He became extremely frightened, but the touching continued—'it was almost like teasing me'" (*Ibid.*, p. 101).

Solid Objects Without Mass

Other physical aspects of UFOs that alert us to their possible spiritual origin are manifested in the accounts discussed earlier. For instance, Ed Walters described his first sighting as though the UFO had no physical effect on its environment. "It glided along without a whisper of sound." In all of the reports of these Gulf Breeze UFOs, the observers are unanimous that none of the UFOs made a sound. "There was no hum, no wind, not a single disturbance to the air, trees, or houses as it passed over them." This description of the UFO having no effect on the wind or air is classic. Large UFOs shoot past observers at tens of thousands of miles per hour, and fail to create a wisp of wind or a sonic boom. Is this possible?—not for an object with mass—not in our space/time continuum. Others report walking on solid, metallic UFO floors, but being unable to make any sound of footsteps. What are we to think of UFOs that change form and structure as abductees watch? Is this the product of superior science? Probably not.

Synchronous Specters

The physical makeup of the "aliens," entities, or beings, is also of great interest from a Satanic-origin viewpoint. We have many testimonies, for instance, that the "small grey" variety often float along without touching the ground, or that they walk and otherwise move in unison. The floating is easily explained in terms of spiritual manifestations. The aspect of moving and walking in "lockstep" unison is somewhat more intriguing. If we were looking at the problem from an extraterrestrial visitor theory viewpoint, we would think that the small greys are in some kind of controlled telepathic communication. That is the initial appearance, anyway. However, why would spiritual beings walk and act in unison?

One aspect of demonic entities that comes to mind is that if a psychically projected scene is being generated by just one of these beings, the movement portions of the manifestations may require more brainpower, or whatever the source of their manifestation abilities, than the being possesses. When Betty Luca was asked to observe a particular scene as an observer, the scene slowed to half speed, apparently due to the "double processing" of the images. Assuming that she was being directly "fed" the images she was experiencing under hypnosis, a subject discussed below, this could explain the phenomenon. As any cartoonist or computer animator can attest, the more movement that is generated in an image, the more work and power that is required. Although we are not on well-trodden ground when discussing these aspects of demonic or others' abilities, we can safely assume that when presenting any manifestation, the beings' forces are taxed in the process—the more glorious or complicated, the more taxing.

I Am Legion

Another characteristic of demonic entities that could explain the "unison" aspect of these apparitions is the "legion" principle. When Christ spoke to the spirit entity that had possessed a young man, asking the entity's name, the entity "answered, saying, My name is Legion: for we are many" (Mark 5:9). After receiving this information Christ listened as the spirits begged him not to send them out of the country: "And all the devils besought him, saying, Send us into the swine, that we may enter into them" (Mark 5:12).

They entered into the *2,000* swine "and the herd ran violently down a steep place into the sea" and drowned. Reserving for a later discussion the overwhelming demonic desire to possess bodies, we well remember the many "alien" encounters in which multiple telepathic voices spoke in unison. For example, speaking of the "visitors'" attempts to control his will and behavior, Whitley Strieber recounts:

> I remembered the visitors' admonition about sweets and decided, experimentally, to toss away the [ice cream] cone. The moment I discarded it, three young voices shouted in unison, "He threw away ice cream for us!"
>
> This sounded totally real, but so close to my ear that it couldn't have been generated by somebody, say, hanging out of an apartment window or standing across the street. I had never heard disembodied voices before the visitor experience started.[5]

Strieber describes how these "disembodied voices" would occasionally speak to him, either within his head or just outside of his ear. He was quite disturbed that he may have developed a brain disease or schizophrenic condition, resulting in his many abduction and disembodied voices experiences, so he underwent thorough psychological and medical testing to determine if he had a diagnosable disorder. He did not.

Betty Andreasson Luca, as well as many others, similarly reported that many of her lifelong contacts with the beings included their speaking in a "chorus" of voices to her. It is important that these beings feigned divinity—speaking in terms of the Savior being "my Son."

> Betty defensively proclaimed her Christian faith . . . . "There is nothing that can make me fear. I have faith in Jesus Christ!"
>
> "We know, child, that you do. That is why you have been chosen. I am sending you back now. Fear not . . . . It is for your own fear that you draw to your body, that causes you to feel these things. I can release you. But you must release yourself of that fear *through my son.*"
>
> The words "through my son" suddenly became the catalyst for the most moving religious experience that I have ever witnessed. Betty's face literally shone with unrestrained joy as tears streamed down her beaming face.
>
> "Oh, praise God, praise God, praise God. (Crying) Thank you, Lord. (Crying, sobbing) I know, I know I am not worthy. Thank you for your son. (Uncontrollable sobbing) Thank you for your son."[6]

---

5 *Transformation,* p. 147.

6 Raymond Fowler, *The Andreasson Affair,* Bantam Books, New York (1980), p. 87.

Eva, another of Dr. Mack's abductees, was forced to allow the aliens to speak through her to Dr. Mack. Notably, she spoke from the "we," and the "we" announced that they were God:

> A shift in Eva's perspective occurred at this point, and for the remainder of the regression she spoke as if from the perspective of the alien community, using only the pronoun "we." "It's like I'm not speaking," she said. The intensity of her experience in this realm was physically difficult for her and caused pain in the hands due to blocked energies. The aliens were gone and she saw the frame of a white triangle. "It's too intense," she said, "they can cause damage" to the human body. The beings emanate "from different dimensions, beyond physical," Eva observed, "and they need somebody that's closer to the human being who is able to communicate somehow physically with them . . . ."
>
> The information the alien beings bring, Eva said, comes from another intelligence, a realm beyond the physical world. But most people disregard it, dismissing it out of fear as "crazy" thoughts or just "imagination." . . . One of the problems in reaching us, she said, is that human beings are anchored in the need for "physical proof" through the "five senses," which "we" are "trying to provide." This is difficult, for "we don't consist of physical data . . . We are not in space/time. We don't have any form . . . We are all. You could say I or one. It does not matter . . . We are an offshoot of I" or "what you would equal to God." Eva herself is "a vehicle through which we can convey that information." As the regression ended, Eva thanked me "for letting us be heard" and felt afraid to open her eyes [*Abduction,* pp. 250-51].

Not only do we witness classic spiritual/corporeal possession of Biblical proportion in these passages, but the "legion" principle is dominant. Dr. Mack, apparently accepting the "aliens" at face value, listens to a female who is frightened of her experience, and rejects it at a personal level, and comments: "Writing in the voice of her cosmic we/she, Eva described the need of the beings to 'adjust our communication from higher vibratory levels to those of earthly (verbal) vibration.' To 'slow down' in this way and 'vibrate at more subtle levels . . . takes training . . . We are using Eva's body with her consent full time now. Earthly Eva has not left, but she has diffused with us so that her earthly powers have been greatly enhanced, so to speak.' She warned of the limitation of using words to describe such profound experiences, especially the relation of the earth plane of existence to other realities" (*Ibid.*, p. 256).

Although we discuss the "possession" principle in greater depth below, the "aliens" begin to reveal their true identity, no matter how much they try to convince us of the great beauty of the Emperor's new clothes.

Ghostly Abilities

Another characteristic of the entities that indicates a spiritual origin is their practice of walking through walls and other solid objects. Their explanation that they are able to change their vibrational levels at will, and pass through solid objects in that state, is not convincing. Although this explanation sounds *possible,* it seems more of a smokescreen to mask a clearly spiritual manifestation or feat than an explanation of superior science or brainpower.

Invasion Of The Body Snatchers

The "disembodied" characteristic to which we have referred in the prior two topics leads us squarely into a third indicator that these are demonic spiritual beings. *They have a need of bodies.* As is clearly manifest throughout the abduction and contactee literature, the procurement of bodies for their species is an overriding element in the motivation of these beings. Their stated reasons for requiring bodies vary with each contactee, but the theme is constant.

In Dr. Mack's sessions with abductees, the purposes behind the abductions were discussed openly between the professionals and the victims. Dr. Mack reports the following conclusion. "For human beings, Pam suggested, having become physical, the body is 'our avenue for learning.' Peter agreed with this 'absolutely.' The beings are like 'God's angels in a very roundabout way,' like 'messengers, or He's working through them, just like He's working through you [Dr. Mack] and me and Pam . . . Pure form consciousness on the spiritual plane does not have the experience of the physical,' Peter observed. 'So it had to embody physicalness in order to have all the experiences of the physical. It's like consciousness said, 'Well, what do I want to learn today?' and it chose the earth as a place to learn physical things'" (*Abduction,* p. 309).

Scratching his head over the implications, Dr. Mack theorizes: "It is as if beings that are semi-embodied and derive from another dimension, have taken on the task (or have been recruited to do so by some other, 'higher,' intelligence) of becoming embodied to the degree that they can combine biologically with human beings. If it is 'genetic' at all in the sense that we know this word, we have no information whatsoever about what the genetic alterations that the aliens create might be which could permit our species to merge" (*Ibid.*, p. 333).

There are three possible methods that the beings could employ to secure these bodies: genetic engineering; mechanical invention; and, corporeal possession.

Theory 1: Genetic Engineering

In the television movie *Intruders,* the "Laura Davis" character comments on the hybrid baby nursery aspect of Kathie Davis's experiences on the UFOs to the "Psychiatrist" character, and queries, "Why do you think they created them—to start a new world?" The Psychiatrist responds pensively, "Maybe—or to save an old one." Although this exchange is a dramatization produced to editorialize on the hybrid baby phenomenon, the question and answer are common in UFOlogy. Abductees report that they "perceive" the answer to be (1) to save the "genetically imperiled" alien race, (2) to save our own "soon to become extinct" race, or (3) to create a new hybrid race, possibly to inhabit a new world somewhere in space.

We have already discussed some of the problems associated with these responses. First, if an extraterrestrial human race, genetically compatible with our own, were to be having such difficulties with their genetic stock, why would they pursue the entire abduction scenario course, when their claimed superior capabilities enable them to engineer such stock or to at least steal the human samples unnoticed? The "aliens" have no intention of working unnoticed! They fly their craft through our dark night skies lit up like neon Christmas trees! They do everything they can to be detected, while pretending to work surreptitiously. They inadequately seal the abduction memory into the subconscious regions of their victims' minds knowing that it will be recalled in a dramatic investigation. The genetic experimentation spectacle is a sham.

Why Hybrids?

Nothing about the "hybrid baby" farce makes the least bit of sense. Why "hybrids" instead of "aliens" if they want to preserve their own species? If this were their true purpose they could use a small amount of human genetic coding for minor physical improvements. If their race is truly superior to ours, which claim they make frequently, why pollute their offspring with inferior genetics? Understanding that, some will think these observations sophomoric, from a genetic engineering viewpoint anyway. I challenge those who accept this theory to explain the necessity of a 50/50 genetic hybrid, instead of a genetically reinforced alien result. It is a smokescreen.

Not To Preserve The Human Form

The second response, "to save our soon-to-become-extinct human race," has no basis in truth. There is little probability of imminent, worldwide destruction of the human race. Despite the cries of the "environmentally zealous," there is no evidence of such calamities. Abductees and contactees report how they are filled with visions of the Earth's imminent destruction. Whitley Strieber speaks of his very vivid induced dream in which he saw a nuclear power plant blow up and the moon explode, causing great catastrophes on Earth.[7] As a result of his "catastrophe" visions (common in UFOlogy) implanted by the entities, Strieber has written best-selling books decrying the use of nuclear energy (a common crusade of contactees) and generally promoting the Earthspirit doctrines of the New Age.

The New Age Earthspirit message is becoming quite intense among abductees. Dr. Mack reports that "Scott's memories moved then into the apocalyptic vision I have heard increasingly from abductees. Major changes in the world are coming. The aliens will only come 'when it's safer.' But that will not occur until there are 'less and less' of us as we die off from disease, especially more communicable forms of AIDS that will reach plague proportions" (*Abduction,* p. 104). Ed, another of Dr. Mack's abductees, is forced to the Earthspirit brink with his cosmic revelations:

> The female being, whose name he now recalled as something like "Ohgeeka" or "Ageeka," picked up on these qualities and underscored the responsibility Ed has for his gifts and powers. "Listen to the earth. Listen to the earth, Ed. You can hear the earth. You can hear the anguish of the spirits. You can hear the wailing cries of the imbalances. It will save you. It will save you . . . Things are going to happen," she said, but he must "listen to the spirits," even if he is taunted, and not feel overwhelmed. "She gave me a flash . . . she opened up that channel and turned up the volume. Some of [the spirits] are crying; some of them are mirthful. She just ran me through the whole thing in a couple of seconds. 'All this you can see, hear, and feel. Other people may think

---

7 *Transformation,* pp. 58-59.

you are crazy.'" The earth itself, the being told him, is enraged at our stupidity, and "the earth's skin is going to swat some bugs off" that do not know how "to work in symbiotic harmony" with it [*Ibid.*, p. 61].

At this point he recalled that he actually saw the spirits in the form of "mirthful little playful creatures, just kind of, just bounding around." I asked him to describe them. "They're like energy forms . . . of many different kinds. There are many different shapes, colors" (he giggled at this point). . . . One of the spirits "is talking to me." He had long, silvery hair and an oversized head and is only a foot or two tall, like a "micromidget." The spirit said, "Well, see, I put myself like this so you could sort of look at me and relate to me. But I don't have to be like this if I don't want to be like this, and I can change myself into a multitude of forms . . . I put myself into this semihumorous shape so that you would feel comfortable and mirthful with me, because I know certain other types of creatures you're afraid of, like spiders, in particular, and snakes. So I'll make myself like this!" [*Ibid.*, p. 62].

In addition to the "innocent play with nature, the way we're supposed to be," Ed remembered seeing "distorted forms of entities, of spirits out here now because man has done so much damage, hurt and damage both to himself and to each other and to Mother Nature." He was shown "grotesque forms . . . Horrific. There's dark, gray, malignant forms that they were trying to heal, or rebalance. It took a lot of effort [of the healthy spirits] to keep these malignant forms from growing in magnitude and malignancy." . . . The malignant, destructive forms, were created by the imbalance of "the collective human psyche . . . dark gray things that just come along and suck up and destroy anything. They just go berserk, and they just zap anything in their path, in their range" [*Ibid.*, p. 63].

Dr. Mack, who is himself full of environmental zeal, is predisposed to accept the "alien" message because of his strong personal beliefs. He is pulled into the phenomenon, and often compromises traditional psychiatric principles in his efforts to justify the exotic claims of the abductees. We see him squirm, however, as he attempts to reconcile the spiritual implications of the environmental message. He reveals that his search for environmental exoneration leads him far from the well-trodden curative path.

Ed's direct encounters during the abduction with shape-shifting spirits that have actual form is interesting, as is the actualization of negative emotional forces in the shape of malignant or demonic spirits. The Western worldview has no place for such entities, tending to regard them as the products of fantasy or the projections of the psyche, although they are believed in or their existence acknowledged by other cultures throughout the world. For example, in a meeting in India in April 1992, in which I participated with a small group of professionals who were invited to discuss the abduction phenomenon with Tibetan leaders, the lamas saw alien beings as spirits that

> have become upset by the invasion and destruction of the earth's environment, which they inhabit as well. A Tibetan physician explained that as a result of our ignorance, material attachment, and aggression, manifested by the desecration of the planet, these spirits have become annoyed and irritated and are causing "negative disturbances." A leading spiritual figure among the Tibetans also saw the aliens as spirits that have become so troubled by human destruction of the realms they inhabit that they have been forced to come among us, seeking our compassion and transformation [*Ibid.*, pp. 66-67].

Any claim to preserve the human race by collecting specimens before the holocaust is untrue—it is not "our race" that is being preserved in the UFO nursery—it is a purported hybrid race. As discussed above, the creation of a hybrid race to repopulate the sterile Earth makes no sense, and contradicts other alien claims of what they are doing.

The Superhumans

What could be the benefit of creating a superior human race then—assuming that such is the case? Two possible scenarios come to mind. The first hypothesis: the only hybrids, assuming that there are any, are those rendered from the altering of human bodies. These would result in the production of lifeless organic shells. These bodies are then made available to unembodied spirits, whose driving compulsion is to possess a body, no matter what its source. The subject has been debated for centuries by speculating theologians who have argued about the sin of copulating with demons:

> The devil does not have a body. Then how does he manage to have intercourse with men and women? How can women have children from such unions? Theologians answer that the devil borrowed [possessed] the corpse of a human being, either male or female, or else he forms with other materials a new body for this purpose . . . .
>
> The devil then is said to proceed in one of two ways. Either he first takes the form of a female succubus and then has intercourse with a man, or else the succubus induces lascivious dreams in a sleeping man and makes use of the resulting "pollution" to allow the devil to perform the second part of the operation.[8]

Many ancient theologians theorized that it is by this very process that Antichrist (or, the Man of Sin, if Antichrist will be a political entity rather than an individual man), must be born, which brings us to our second possible scenario for the creation, actual or pretended, of a superior human race.

The second hypothesis: The entire New Age movement is based on the belief that mankind is about to evolve into a new, improved, superior, enlightened race. The elite of the New Age are believed to be made up of "early evolving" ones of the coming superhumans. Their superhuman abilities—intellectual, spiritual, and physical—will enable them to take the reigns of government and religion, and save the world from the imminent catastrophes that threaten its certain destruction. They will be supported by the "space brothers" and "ascended masters" who are concerned for the welfare of our planet and its inhabitants. Anyone wishing to receive the benefits of their leadership must abandon his outdated, superstitious beliefs, and embrace the New Age. Antichrist will thus come to power with the aid of these men.

Lewis and Shreckhise, Fundamentalist Christian New Age watchdogs who adhere to this UFO/Satan theory speculate:

> If our calculations are right there will be a massive open manifestation of alien presence and power on Earth. Something big is about to happen! Kings, prime ministers, and presidents will tremble before the "ascended masters." How could they refuse the help so generously offered when Earth is about to self-destruct? They will play right into the hands of Antichrist if they follow the aliens' suggestions. Some rulers may be executed by angry mobs for taking a stand against the "space brothers."
>
> An axis-shift and concurrent disasters are related to the "paradigm shift" that New Age devotees expect in the near future. This shift in humanity's worldview will be precipitated by massive demonic deception and even possession.[9]

The second reason for creating a seemingly superhuman race, then, is to convince humanity to follow the superhumans into whatever paths the superhumans may choose. Whether these superhumans are merely humans with supernatural powers, hybrid-appearing humans endowed with supernatural powers, hybrid-appearing demons, or human-appearing demons, the predicted result is the same. How likely is it that the superhuman race theory is true? In one of the four forms just enumerated, the scriptures support the probability of such a ruse—in at least partial implementation of Satan's last-days, premillennial rise to world domination. The Man of Sin, the head of Antichrist, will be such a man, according to the Bible.

> [The man of sin] opposeth and exalteth himself above all that is called God, or that is worshipped; so that he as God sitteth in the temple of God, *shewing himself that he*

8 *Dimensions,* p. 146.

9 *UFO: End-Time Delusion,* p. 160.

> *is God* . . . . Even him, whose coming is after the working of Satan with *all power and signs and lying wonders,* And with all deceivableness of unrighteousness in them that perish; because they received not the love of the truth, that they might be saved. And for this cause *God shall send them strong delusion,* that they should believe a lie. [2 Thessalonians 2:3-11; emphasis supplied]

Satanic Science

Are the beings carrying out genetic engineering experiments for any of the reasons discussed above? There exists a low likelihood that extraterrestrial humans are involved in such an activity for the reasons allegedly proffered by them. The question of Satanic involvement in such physical pursuits is intriguing, and opens the floodgates to many questions involving "Satanic Science." We, of course, do not know to what extent Satan may involve himself in human reproductive activities to promote his diabolic purposes on Earth. Furthermore, we know even less about his ability to enter and possess such custom-engineered flesh. History and scriptures shed little light.

One example that we have of spiritual intervention in human insemination is the conception of Jesus's body. When Mary was told by the angel Gabriel, "And, behold, thou shalt conceive in thy womb, and bring forth a son, and shalt call his name JESUS," she responded, "How shall this be, seeing I know not a man?" (Luke 1:31, 34) Gabriel then explained, "The Holy Ghost shall come upon thee, and the power of the Highest shall overshadow thee: therefore also that holy thing which shall be born of thee shall be called the Son of God" (Luke 1:35) This "explanation" has spawned a plethora of theories and debates concerning the nature and origin of Christ Himself.

Modern fertilization techniques could enlighten us as to how it occurred. We are now capable of taking physical seed from a man and artificially implanting it in the ovum of a woman. This should not be too arduous a task for the Holy Ghost—to take a small seed from Deity and place it in the ovum of a mortal woman. So doing, all of the known facts regarding the conception of Christ are undisturbed. Although this conclusion leaves room in the world for numerous virgin births, there is only one Only Begotten Son of the Father.

Having considered the uniqueness of the conception of the Savior, it is more than curious that much of the activity described on UFOs is centered in artificial insemination and the premature removal of hybrid fetuses from artificially inseminated human women. What drama is being played out here? A counterfeit of the virgin birth?—a probable claim that will be made by the "superhumans," in all of their glorious modesty.

This only tells us that at some level, angelic or higher spiritual beings have the power to intervene in human physical genetics and reproduction. Can lower forms of spirit entities do the same? We can only speculate. However, at the center of the alien encounter/human abduction phenomenon, which has become paramount, we find one constant—the beings need access to human bodies because of some problem with their own "physicalness." Many abductees have described the beings in the same terms as Dr. Mack's Arthur: "I asked Arthur to describe the beings. They looked 'like an embryo,' he said, 'just little gentle things.' They were 'luminescent,' and semitransparent, with large heads and small bodies, 'little skinny arms and little fingers. Maybe not five fingers.' In addition, he saw 'little legs,' smooth faces with 'a little mouth, little noses' and 'no hair . . . like babies, like embryos,' he said again. The eyes were all black and rounder than ours" (*Abduction,* pp. 378-79).

Some commentators have noted that the embryonic appearance of the beings suggests that their bodies are being cloned from human genetic material, or are being quick-grown from aborted or abducted human fetuses. Either way, this is a very disturbing possibility, and only strengthens the case for demonic possession of genetically or otherwise manufactured bodies.

Theory 2: Mechanical Invention

As mentioned above, there exists a little-discussed subject that must be raised here in an effort to determine if the "aliens" are of a mechanical nature, or of some other "manipulation and restructuring of matter" as also mentioned above—is there such a thing as Satanic Science? God inspires and metes out technology in His own time, for His own purposes. Does Satan do the same? Surely, any technology that is known to mankind is also known to Satan. He has this kind of knowledge, and uses it to his own advantage every time he sees an opportunity—just look at how he has used film and television for his own purposes. Possibly, Satan is aware of even greater technology than is dreamed of on Earth, he having filled a high position before his rebellion and fall. Our technology is simple and crude. The microprocessor and wave technology are the only real advancements made in centuries. Before that, the light bulb—not very impressive. Our question is then: Does Satan conduct technological enterprise and/or dispense technology to his human recruits for his own purposes?

Because we were unable to answer this question,—does God personally use technology?—for similar reasons we may be equally unable to answer whether Satan does. However, we assume that God dispenses it to man and we have no indication that Satan cannot. The only force that could prevent Satan's employment or dissemination of technology for his purposes is God's disallowance. Would God disallow it? We know that he puts many limitations on Satan to control the level of his attack on humanity. Otherwise, humans would be in constant bondage to demons.

Is Satan involved in creating robotic or materialistic bodies for him and his demons to possess and control? We know that Satan can enter and control the bodies of animals, and that he can influence objects with mass. Therefore, little reason exists to doubt his ability to control mechanical devices, including sophisticated robots, as though they were alive. In fact, some of our most eerie accounts of supernatural displays are centered in just such acts of animating objects.

Is this the answer then to UFOs or their humanoid occupants? Within this theoretical line of reasoning two possibilities exist. First, that Satan creates physical crafts and bodies, then animates them through spiritual control, like possession. Second, Satan recruits humans to his cause, arming them with technology sufficient to build and pilot the craft, and to build and control the robots, if such are employed under this theory.

It is often reported that the small greys and other "aliens" are seen moving in a very mechanical, robotic fashion. This could also explain the phenomenon of why they are seen moving in unison, even when the movement of the second entity has no effect on the object being manipulated by the first. Whether these fabrications are mechanical (made by Satan or human recruits) or illusion, the operation and results are the same—they are centrally controlled. This would also explain why communication with the beings is telepathic—they do not have the power of human speech.

The fact that the smaller, gnomelike, or robotlike beings are often seen in the presence of, and are subservient to, human-looking beings supports these theories. Are the human-appearing entities behind the UFO phenomenon Satan's human recruits who are employed in his service? Are they demonic beings projecting an angelic or human appearance? Are they also robots—controlled spiritually or mechanically? We, of course, have no firm answers. But we know for sure that none of these "message bearing" beings is what it claims to be. They are all lying—a prime indicator that we are dealing with Satan.

Whitley Strieber had an experience that lends credibility to our discussion. After being embarrassed by the beings in a bizarre drama, he recounts:

> I was forced by my [ballooning paper gown] clothing to move like an arrogant prince—which made me feel even more like a toad. Carrying myself as best as I was able, I left the room. We were going down the curving corridor again when one of the blue beings looked up at me with his wide face. I saw it clearly this time, and it was really startlingly horrible. Awful! The eyes glittered as if they were shiny black membranes, with something moving behind them that made lumps and pits as it seethed within the eyeball. He smiled, showing the tips of his grey, spongy-looking teeth. His companion pulled open one of the [many] drawers [that lined the corridors].
>
> *In that drawer were stacks of bodies like their own, all encased in what looked like cellophane.* Their eyes were open, their mouths wide as if with surprise. I did not know what to make of it. The oddest thing was the way the drawer was opened with a prideful flourish. I was being shown something the two of them clearly thought was wonderful.[10]

An arsenal of readied bodies. Ready for what? For whatever it was that propelled the two gnomes that so proudly exhibited the contents of one of the many drawers. If these were manufactured bodies, why these hideous little gnome appearing creatures instead of handsome humans or angelic beings? After all, if you are going to create your own tabernacle through some kind of technology or manipulation of matter, it is just as easy to generate a Lamborghini as a Pinto. The answer lies in the fact that this is a drama—a theatrical production—and the actors are costumed appropriately for their parts. If Streiber's experience was indeed of a physical nature rather than an illusion, it is apparent that these inanimate bodies he saw were lifeless, just as those of the escorts had been prior to animation.

---

10 *Transformation,* p. 40: emphasis supplied.

Similar Encounters Have Historical Precedents

In analyzing the beings and their behavior, some researchers have shown the parallels between human encounters with nonhuman beings throughout history and these UFO encounters. Noted scientist and UFO researcher Jacques Vallee has produced a wealth of information concerning these parallels. Some of them are included earlier as paranormal parallels.

One such encounter is cited by both Whitley Strieber and Jacques Vallee, albeit for different reasons. The tale is well documented and relates how a young teenage boy disappeared for two years. One day his mother, who had grievously mourned his death, was shocked to see him at the door. Upon seeing him she was astonished that he had not aged at all during this period of traditionally accelerated maturation. Vallee cites the story because, as in many such cases, the boy insisted he had been away with the little people for less than two full days. Strieber, however, found interesting the fact that the boy carried a bundle containing a white, seamless paper gown, exactly like the one given to Strieber in the above-quoted story, that the boy told his mother the little men had given him to wear while he was with them.

Following is a story quoted by a researcher cited by Vallee to further demonstrate the similarities between traditional encounters with the "Gentry" and modern UFO occupant encounters.

> The folk are the grandest I have ever seen. They are far superior to us and that is why they call themselves the Gentry. They are not a working-class, but a military-aristocratic class, tall and noble-appearing. They are a distinct race between our race and that of spirits, as they have told me. Their qualifications are tremendous: "We could cut off half the human race, but would not," they said, "for we are expecting salvation."
>
> They take young and intelligent people who are interesting. They take the whole body and soul, transmuting the body to a body like their own.
>
> I asked them once if they ever died and they said, no; "We are always kept young." Once they take you and you taste food of their palace you cannot come back. They never taste anything salt, but eat fresh meat and drink pure water. They marry and have children. And one of them could marry a good and pure mortal.
>
> They are able to appear in different forms. One once appeared to me and seemed only four feet high, and stoutly built. He said, "I am bigger than I appear to you now. We can make the old young, the big small, and the small big.[11]

11 *Dimensions,* p. 54.

It is difficult to differentiate between traditional elf tales, and modern alien encounters. To further illustrate the similarities Vallee says:

> The parallel between these modern claims and the medieval legends is closer than ever. The same theory was presented about intercourse with the elves. I have shown that fairy tales are full of stories about the stealing of human babies, changelings, and the abduction of both males and females for procreation with the Gentry. Even the scars sound familiar.[12]

The entities' assertion is quite interesting: "We could cut off half the human race, but would not, for we are expecting salvation." Again, the number two billion comes to mind. Interestingly, the Gentry describe themselves as "a distinct race between our race and that of spirits."

By this time it should come as no surprise that for centuries clergymen and town officials have documented the testimonies of thousands who have recounted tales of abduction, procreative experimentation, and other encounters with nonhuman beings, which closely resemble modern accounts of UFO close encounters. Top thinkers in UFO research, including Vallee and this author, believe that the phenomena spring from a common source—whatever it is. Whitley Strieber writes in his forward to *Dimensions:*

> [Vallee] reveals an appalling truth: the phenomenon has been with us throughout history, and never have we been able to deal sensibly with it. Whatever it is, it changes with our ability to perceive it. The fifteenth century saw the visitors as fairies. The tenth century saw them as sylphs. The Romans saw them as wood nymphs and sprites. And so it goes, back into time. [*Forward,* pp. vi-vii]

As we have seen, the procurement of bodies is an underlying compulsion of the beings. Are these genetic experimentations, hybrid children, and inventoried bodies all part of Satanic Science, providing what God has denied the satanic beings—mortal tabernacles? Are Satanic forces, demonic or human, literally manufacturing physical bodies for demonic habitation and animation?—or, are these characteristics of the UFO phenomenon just more smoke and mirrors to create an illusion for their audience? Again, we cannot know the answers to these questions absent more information. One thing we know for certain—from this time forward we cannot rely solely on our five physical senses to discern what we think we are seeing or hearing.

---

12 *Ibid.*, at 267.

Theory 3: Corporeal Possession

Demonic infestation and possession of humans is well documented throughout the entire history of the world. Again, this results from numerous demonic spirits' overwhelming drive to unite with the elements and "feel" the physical world—it is the only means by which the sensual lusts of depraved, unembodied beings can attain any degree of satisfaction. We see activity in this arena escalating as we approach the winding up scene. Spiritual merging with these entities is the affirmative pursuit of most Eastern and New Age religions and cults. Even unwilling, righteous people have become plagued by the infestation of such vile spirits. Does "demonic possession" play a role in the UFO phenomenon? To at least some degree, the response is a clear "yes," as we have already seen.

New Age guru Ruth Montgomery explains that the Earth is being visited by foreign beings:

> To the limit of my understanding, it would seem that the Guides have identified three types of beings who have achieved sufficient advancement to enter our earth plane, and appear in solid form to us . . . .
>
> Apparently the highest achievers in this category are the avatars, who can come and go at will, and who, according to the Guides, are in touch with outerspace beings as well as with humans on Planet Earth. The second type is the Walk-ins, who have always been Earthlings but are high-minded, advanced souls who return to adult bodies in order to accelerate the progress of their fellow-men. The third class is the extra-terrestrials, who, still few, have allegedly found the means of penetrating the earth's atmosphere and occupying bodies of adult humans, for limited scientific experiments and observations of our planetary changes.[13]

Montgomery further clarifies on another occasion:

> I asked about the different means used by space beings to enter our culture. and the Guides said that some are being born into human bodies for the first time, some have lived here upon occasion before, some are arriving as Walk-ins, and others are temporarily exchanging bodies with Earthlings "with or without permission."[14]

We must remember while reading these New Age writings that Ms. Montgomery is not apologizing or making excuses—corporeal possession is a common occurrence and completely accepted by New Agers as a nat-

---

13 *Strangers Among Us,* pp. 147-48.

14 *Aliens Among Us,* p. 13.

ural means of spiritual communication and interaction. Channelers like Montgomery receive the above tidbits of information frequently from their "spirit guides," spiritual entities who profess to be advanced, enlightened beings, but who are indeed, the very same demonic spirits that serve Satan.

This self-admitted practice of entering the bodies of Earthlings "with or without permission," to carry out extraterrestrial activities is a strong indicator that corporeal possession is pursued at some level. And whether the assertion is true or not, we at least know that the "extraterrestrial" contacts feeding us this information are demonic spirits.

New Age author Brad Steiger quotes the words of a channeled extraterrestrial who explains the process in more innocuous terms: "By mutual agreement between a planetary dweller and an inhabitant of our craft, the knowledge and the memory of one of us may be blended with the planetary inhabitant without the loss of the receiver's identity."[15] However, in assessing the communications and New Age gospel message being delivered by the UFO occupants, Steiger is hesitant to accept the entities at full face value. "I had the uneasy feeling that the ecstatic flame may, in reality, have been kindled by multidimensional beings who have a kind of symbiotic relationship with man and who may exploit their 'prophets' for selfish, parasitical purposes."[16]

Brad Steiger is correct. These "multidimensional" beings are not our friends. They are our enemies. They are deceiving us with their extraterrestrial guise, and are enjoying greater success than at any prior time in duping humans into believing a great lie. Even the best scientific minds investigating the phenomenon see that classic spiritual possession is occurring, but are so enamored of the Emperor's new clothes that they refuse to see the grand illusion for what it really is. Without realizing it, Dr. Mack supplies us with the most complete data to date that evidences the "possession" of abductees by their abductors:

> At this point in the session the timbre of Peter's voice changed to a kind of monotonous droning and he shifted to speaking from the alien perspective. "We," he said, want to study the chemical reactions of the brain, and how people will react in order to "know when it is time to be present . . . For as we measure the impulses," the voice continued, "we will know at what level the shock will come in, so we will be better able to control it so we will be in tune with the [human] beings as they go

---

15 *Revelation,* p. 141.
16 *Ibid.,* p. 8.

> through the shock process, as they go through the unfolding of seeing us for the first time. . . ." After the regression we spoke of the channeling of the voice that had come through during the session. He recalled that this had begun four years ago in the Caribbean, and that it feels like an expansion of his own energy which allows the "alien energy" to come through him. This can occur, he said, when he can "surrender my mind, surrender my ego." He (like other abductees who have this capacity—see, for example, Eva and Jerry) does not altogether trust the information he receives in this way, but does feel that "it's coming from a higher consciousness, from the spiritual plane." It was occurring in present time, he said. "It's live broadcasting" [*Abduction,* pp. 303-04].

Like millions of American adults (according to the Roper Organization survey) Dr. Mack's subject, Sheila, "also began to have recurring dreams in which she would experience terror, be unable to move, and her body would feel as if it were vibrating or 'full of electricity.' At first she called these 'spiritual dreams,' and they made her feel like someone or something else were controlling her body, as if she were 'possessed' by demons" (*Ibid.*, p. 70). These spiritual feelings are echoed by Joe:

> In the third regression Joe had the intense experience of feeling his alien and human parts come together, a profound, ecstatic expansion, a kind of rite of passage, that contained both terrifying and joyous sequelae which extended and deepened the process in the weeks that followed. Joe's experiences, especially those related to Mark's birth, demonstrate dramatically the separation or discontinuity of his being or soul from his body. The lightness of the soul's experience in the "spirit" or "other" realm—our language fails us here—contrasts with the density of the physical body as experienced in the earth domain.

Dr. Mack comments, "The aliens are experienced by Joe and many other abductees as much more closely connected with the divine source [God], or *anima mundi,* than are human beings, who are struggling to overcome their extreme separation. Therefore, the coming together of these dimensions of himself brings about, virtually by definition, Joe's own deeper connection with the divinity, a sense of oneness with all beings—essentially his reensoulment. *Curiously, the alien beings seem, conversely, to long for a deeper connection with humans, as if the greater density of our embodiment or physicality contains some sort of appeal to them*" (*Ibid.*, p. 198, emphasis supplied).

Unknowingly, Dr. Mack sets out quite well the historical relationship between humans and the fallen one third of the hosts of heaven. As a footnote to Joe's experience, Dr. Mack adds, "Four days after the session he had a kind of energy crisis or 'kriya,' evoked during a massage. He sweated and shivered and felt intense pain moving from place to place in his body, starting in the kidney region and flooding his spine and head. 'I was moaning and rolling around, overwhelmed by the experiential/emotional/physical pain.' His 'ET guides' were holding his hands and head, and he was flooded with scenes from his past . . . ." (*Ibid.*, p. 191).

The result of all of this indoctrination and brainwashing by the "aliens" is astounding. The abductees are not only beginning to believe the false message, which on its face strikes them as absurd, but they are doubting themselves and their own individual identities, as we have also discussed before. Scott, another Mack abductee, speaking in his alien identity, writes to Mack: "I fear humans more than anything else. We have tried to change you many times. Many members of our species have been destroyed in the process . . . I must say the human being has very heightened emotions, too much for me to process at times. We are very sensitive, but our emotions are not as primitive as your own." At this point Scott exposes a very revealing aspect of the "aliens"—"*Your emotions are recreation in a sense. We are happy to be able to feel more than we normally feel. Our fascination [with humans] revolves on this. Our evolutionary process has deemed emotions less important than understanding, but it's like candy to a child your emotions to us. It is like a drug that we enjoy very much*" (*Ibid.*, p. 107).

In his summation of the abduction phenomenon, and noting that Scott's revelation is the viewpoint held by most of his abductees, Mack reasons, "The aliens, on the other hand, long to experience the intense emotionality that comes with our full embodiment. They are fascinated with our sensuality, our warmth, our capacity for eroticism, and deep parental affection, and they seem to respond to openhearted love. They act at times like love-starved children. They delight in watching humans in all sorts of acts of love, which they may even stage as they stand around watching and chattering as the abductees perform them" (*Ibid.*, p. 416).

Mystic Knowledge

The spiritual "knowledge" passed from the UFO entities to "planetary dwellers" bears a strong resemblance to the mystical knowledge with which Eastern initiates are endowed at the pinnacle of their spiritual quest. As discussed earlier, the transference of such spiritual knowledge is another counterfeit employed by Satan to mock the true principle of the ministering of the Holy Ghost. As we know from the Bible, the Holy Ghost radiates His influence throughout the universe on an ongoing basis. However, on special occasions He personally merges with the spirit of a worthy person and shares eternal truth at a level that transcends any worldly communication. The recipient is endowed with spiritual truth that is unmistakable and undeniable. The Bible is filled with examples of this kind of direct spiritual communication between beings sent from God and human recipients of these full sensory messages—the prophets and apostles.

"Psychic implanting" is the means by which Satanic forces convince people that they have experienced past lives or other mystical phenomena. In stating their conclusion about the origin of UFOs and their deceptive purposes, Lewis and Shreckhise declare their belief that psychic implanting is the same method used to convince percipients that they have seen UFOs and their occupants.

> The spirituality that the New Age/UFO religion offers is a false hope. It is based upon subjective human feelings and observations that are being manipulated by demonic forces. The psychic implantation of deceptive experiences is changing the way [abductees] perceive reality. What they experience is a real experience—deceptive and manipulative in character. It is not, however, an experience of reality. It is an illusion of reality used solely for the purpose of convincing people to turn away from the Creator.[17]

Of course, even if we accept the fact that UFOs are a Satanically generated phenomenon, the psychic implant theory is not as amenable to photography and radar detection as are materialization and manifestation generation. Of course, the psychic implant is very amenable to the more peripheral aspects such as alien encounters and communications and the attendant visions of alien worlds or UFO interiors. Commentators suspect that a mixture of techniques is employed to create a panorama of special effects in a UFO/abduction experience.

Whitley Strieber recounts how the aliens touched his head with a "wand," producing realistic, three-dimensional images that swirled about in his mind, reminiscent of the images of dogs and naked women that were thrust

---

17 *UFO: End-Time Delusion,* p. 235.

into the mind of Ed Walters during an "attempted" abduction. The visitors would communicate with Strieber (and countless others) any time they wished. He periodically would hear voices speaking to him as if standing next to him or within his head that would give him information and instructions on what was going on around him. This all sounds very much like psychic implanting. Was Strieber "possessed"? He worried about this possibility, saying "I realized that they must have been aware of my attempts to walk in the woods. I felt more than watched; I felt entered and observed from within."[18] Did he feel menaced? Let's reexamine his fearful wondering.

> Increasingly I felt as if I were entering a struggle that might be even more than life-or-death. It might be a struggle for my soul, my essence, or whatever part of me might have reference to the eternal.
>
> There are worse things than death, I suspected. And I was beginning to get the distinct impression one of them had taken an interest in me.
>
> So far the word *demon* had never been spoken among the scientists and doctors who were working with me. And why should it have been? We were beyond such things. We were a group of atheists and agnostics, far too sophisticated to be concerned with such archaic ideas of demons and angels.
>
> Alone at night I worried about the legendary cunning of demons.[19]

Strieber worried:

> I could not shake the idea of the soul predator. I took my midnight walks regularly now, and every time I reached the darkest part of the woods the thought would come whispering back.
>
> I had no evidence that it was true. I just couldn't rid myself of the notion that there was something predatory about the visitors. I had terrifying memories of them—memories of leering visitor faces, of long, four-fingered hands, of recoiling at their touch.
>
> Those moments remain as if sealed behind smokey glass. I couldn't tell where the memories came from. There were dozens, maybe hundreds of them [*Ibid.*, p. 176].

It is interesting that Strieber reached a point in his hypnotic regression therapy at which he felt that he could no longer trust these memories whose origins he had come to doubt. He knew the memories were of exter-

---

18 *Transformation,*, p. 136.
19 *Ibid.*, pp. 44-45.

nal origin, but doubted their veracity—as if they were a "mask" of something else that had actually occurred.

Strieber too had the clear impression that the "visitors," as he called them, were evil—unnatural.

> In the wee hours of the night I abruptly woke up. There was somebody quite close to the bed, but the room seemed so unnaturally dark that I couldn't see much at all. I caught a glimpse of someone crouching just behind my bedside table. I could see by the huge, dark eyes who it was.
>
> I felt an absolutely indescribable sense of menace. It was hell on earth to be there, and yet I couldn't move, couldn't cry out, couldn't get away. I lay as still as death, suffering inner agonies. Whatever was there seemed so monstrously ugly, so filthy and dark and sinister, of course they were demons. They had to be. And they were here and I couldn't get away I couldn't save my poor family.
>
> I still remember that thing crouching there, so terribly ugly, its arms and legs like the limbs of a great insect, its eyes glaring at me.
>
> And there was also the love. I felt mothered. Caressed. Then the terrible insect rose up beside the bed like some huge, predatory spider. The eyes glittered as it tilted its head from side to side.
>
> Every muscle in my body was stiff to the point of breaking. I ached. My stomach felt as if it had been stuffed with molten lead. I could hardly breathe.
>
> The next thing I knew, something had been laid against my forehead. I felt it there, a light electric pressure vibrating softly between my eyes.
>
> Instantly I seemed to be transported to another place, a stone floor with a low stone table in the middle of it. The table was a bit more than waist high and on it there was a set of iron shackles. A man was led down some steps and attached to these shackles. He was right in front of my face, not two feet from me, looking directly at me with eyes so sad that I almost couldn't bear it [*Ibid.*, pp. 181-82].

Whose Will?

At that point the holographic (or otherwise projected) man was tortured for Whitley's lack of obedience to the commands and requirements of the visitors. When returned to his bed, Strieber was also forced to listen to his son's screams downstairs as punishment for his disobedience.

These forms of psychological abuse to supplant the will of the victim for their own will is reminiscent of Satan's plan to direct the affairs of the Earth through coercion. This, of course, is in direct opposition to God's Plan in which man chooses good or evil of his own free will. Remember Betty Luca's report that the beings were perplexed and obstructed by man's free will. It seems to be a constant theme that runs through the UFO phenomenon.

Strieber describes being raped by an "alluring, yet despicable-looking" alien or hybrid female. Horrified, he screamed out, "You have no right to do this to me. I am a human being!" The female sternly replied, "We do have a right!" This single, affirmative claim of the visitors has perplexed researchers. It would be one thing to conduct experimentations on lower lifeforms with neutral moral intentions, much as we do with our own medical research or migration/habitat tracking. It is another thing altogether, however, to positively assert that one has an affirmative *right* over another creature. This would imply "ownership," which is an aspect of the visitor experience heretofore undiscussed.

One night, Strieber felt that the aliens were returning for him. Again we read how he tried to resist, but he finally lost his strength and became resigned to his fate.

> There was no question of my doing anything about the fact that I knew the visitors were here. It was all I could do to climb the stairs to the bedroom . . . . I felt an absolutely indescribable sense of menace. It was hell on Earth to be there, and yet I could still not move, couldn't cry out, couldn't get away. I lay as still as death, suffering inner agonies . . . . I thought I was going to suffocate. My throat was closed, my eyes swimming with tears. The sense of being infested was powerful and awful. . . . [20]

As we read through Whitley Strieber's accounts of his encounters with the visitors, we wonder if he hasn't possibly employed his proven ability to research and write on unusual subjects. After all, he earned more money for *Communion* than most of us earn in a lifetime. Strieber concludes *Transformation* with his testimonial:

> Of course, one could take the comfortable road and say that I am lying, that the descriptions in my book are hyperbole or hallucination. But they are not. I am telling the truth of what happened to me, and the implications are there for anyone to see. Not only are we not alone, we have a life in another form—and it is on that level of reality that the visitors are primarily present.
>
> I call them visitors, but now I am beginning to think that is a misnomer. I have had the impression that they think of themselves as family, and perhaps that is exactly what they are [*Ibid.*, p. 201].

Dozens of people have added their testimonies to support Strieber's story—they claim that they too saw many strange events at Strieber's remote cabin. An officer with the publisher of *Communion* relates that he

---

20 Whitley Strieber, *Transformation,* pp. 189-92.

was essentially accosted by two hybrids in obvious disguises at a bookstore as he was passing by the new *Communion* display. The book had just been put out, and had not been available to the public—no one even knew that the famed fiction writer had produced a nonfiction expose´ of the visitors. The two characters, having no apparent way of knowing that this fellow passing through the store represented the publisher, flipped through the pages in just seconds repeating "that's not how it was, that's not how it happened at all." In the Whitley Strieber case, the theatrics of the entities knows no bounds. Why should the visitors go to these lengths to support Strieber's incredible story? Because they want us to believe every incredible word.

Sara, a Dr. Mack abductee, was being coerced by the beings to write things down that they told her "in her head." She had been in a serious car accident a few days before she was scheduled to meet with Mack due to "alien" influence on the driver. Dr. Mack reports:

> When Sara would shut her eyes after this accident, in addition to seeing the "light beings," she could also differentiate a second type of entity. When "I shut my eyes I see them . . . I see these guys . . . down in a little row, like three or four little dark guys. Like gibbering." Later she said, "it seems to me like these guys are in my head." In contrast to the "light beings," she described the "other ones" as "frenetic." Shortly after the accident, she felt compelled to do her "listening" every day and to write down the information she obtained. She felt this would render additional accidents unnecessary.
>
> A few days after the accident, Sara and Miguel had an experience in which an unexplained green/yellow light penetrated their room. Miguel is ordinarily fearless, Sara said, but they were both terrified, and he appeared to have lost consciousness for a time during the incident. Sara felt as if she were "physically pinned down" and unable to move. She saw "three things hovering above me" like "three shrouded heads," and thought to herself "something like, listen, we're communicating. This is for real. Something like that, like, get your act together and start writing it down." Then "the whole thing kind of dissipated" [*Abduction,* pp. 205-06].

This confusion over identity, alliances, and free agency (or self-directed choice) is well documented in the case of Peter. "He wondered, 'Who am I?' and felt he was 'losing myself to some degree . . . I am an individual soul that just came into this lifetime' and entered 'Peter's body.' In other lifetimes, past and future, he would be 'something else,' but 'I was always Peter. I'm always this soul . . . My soul is actually alien . . . I'm not sure what that actually means,' he said, whether he was 'created by them or them coming into his body.' If that was the case, 'then I feel lost, like, who am I? Am I just a vehicle for their use? Am I just a machine now? How much of this is my free will? It puts a whole other twist on it'" (*Ibid.*, p. 325.)

Indeed. It puts a whole other twist on it. The goal of possessing human bodies with foreign spirits is now coupled with robbing humans of their free will, either by force or deception. Demonstrating this point further, Dr. Mack asked Paul how he was able to get certain information out of the beings, and Paul's response was quite revealing: "I don't know," Paul replied. "I just, like, let go. I just keep letting go, I guess, which is a big thing I do in accepting information" (*Ibid.*, p. 236). Just keep letting go of what? Of himself, of his own free will to act and decide for himself.

Scott's "surrender" is perhaps the most informative detail in Dr. Mack's research into the abduction phenomenon. Dr. Mack relates, "As has occurred in the case of several abductees with whom I have been working recently, Scott's full acknowledgement of the reality of the alien presence has led him to the realization that he has always had a kind of dual identity, and is capable of experiencing himself as both human and alien. The alien perspective, which apparently has always been imbedded in his consciousness, was not available to him until he surrendered the illusion of control" (*Ibid.*, p. 108)

Is this the true alien message? Are we being told collectively to do what the abductees must do individually in order to survive the intolerable? Must we "surrender the illusion of control" in order to qualify for fitness for the next step in human evolution—the quantum leap into the next, select phase of existence? Whether or not we accept it, this is the message.

Strieber's comment, "I have had the impression that they think of themselves as family," is noteworthy. These bizarre beings who communicate through theater and abuse their victims with the most vile forms of psychology have a purpose behind their eccentricity—could it be that they still think of themselves as our brothers and sisters? Are they truly still truly still expecting salvation? Possibly they believe they can get a second chance if they can succeed in destroying our chances of fulfilling God's Plan.

Many Indicators Of Corporeal Possession

When analyzing the "message" given to Betty Luca for later delivery to the world, it is important to first consider the method of its delivery—first to Betty, then to the world. As discussed previously, except for a couple of brief anomalous experiences, Betty, like most abductees, had no conscious memory of her abductions. It was only as she underwent regressive hypnosis that she was able to "recall" her experiences. Raymond Fowler emphasizes Betty's exceptional ability to summon her lost memories under hypnosis:

> Under hypnosis, Betty not only recounts but *relives* her UFO encounters. She does this in intricate detail—with corresponding emotion, trauma, and body movements. She is able to provide detailed drawings that tally exactly with her verbal testimony. Weeks, months, and even years later she is able to relive selected segments of her experience upon demand by a hypnotist.[21]

Not too surprisingly, many of the abductees who have undergone similar experiences and investigations have such detailed recall to the point of "reliving" the experiences, including Kathy Davis and Bob Luca. Budd Hopkins claims that physical manifestations such as hypothermia (abductees often complain of being in very cold environments) return as the abductee relives the experience under hypnosis. He always has warmed blankets on hand during such hypnotic sessions. One might easily dismiss these occurrences as mere psychosomatic manifestations of a perceived experience, but the phenomenon appears to go beyond that, as related in the following:

> The aliens placed Betty on a soft, rubbery, cushion-like mat on the floor of a section of the craft that was roofed by a large transparent dome. A mouthpiece was installed that kept her tongue held down. When describing it to us under hypnosis, she actually talked as if something were holding her tongue down. Betty began sinking into the rotating circular mat as the craft accelerated upward. Incredible as it may seem, her body actually sunk into the hypnotist's chair! The psychosomatic effects on Betty's face and voice were fantastic to behold. All present were amazed to actually see the effect of the g-forces on her face. The skin got very tight around her face and her mouth was pulled way back. She experienced difficulty talking.[22]

---

21 *The Watchers,* p. 353.

22 *Ibid.,* p. 10.

These manifestations of physical forces on the body of Betty Luca and others during hypnosis give us great pause. No answer exists in science to explain Betty's sinking deeper into the hypnotist's chair as she experienced a mere "recall" of rapid upward acceleration. This is clearly not within the realm of psychosomatic reaction. Some outside force had to be present to compress Betty's body into the cushion of the chair. The researchers present assumed that there exists a latent effect originating with the experience, which remanifests itself as the abductee relives the experience under hypnosis—an effect completely unknown to us. This does not seem like a workable solution.

There is every indication that spiritual or demonic forces are at work on Betty during her hypnosis. Is this why she, like others, becomes very cold at times during hypnosis—is this a "cold spot" manifestation? Is this why no memory exists prior to the hypnotic session—because the memory is directly "downloaded" to her during the session? We know that it is often during these hypnotic trances that people with no prior recollection of abnormal memories are suddenly deluged with vivid, panoramic recollections of past lives. Others are known to levitate during hypnotic trance. These paranormal experiences are common occurrences in both UFO encounters and the occult.

Or, does hypnosis merely unlock or reinitiate an already experienced demonic encounter? Some victims, Whitley Strieber for instance, feel that the encounter really occurred remotely, and it was masked or altered in the subconscious. That is why, they believe, some people who have anomalous, vague memories of animals (generally large-eyed animals like deer or owls), are recalling a masked memory instead of the actual visitor encounter. Deeper than this, some believe that the entity encounter recovered memory is only a mask for a more profound experience—one that the entities really do not want the victim to remember.

Hypnosis—A Spiritual Revolving Door?

The entire process of hypnotic retrieval of suppressed memories is suspect from a spiritual viewpoint. It is commonly held by many Christian churches that certain barriers to spiritual intrusion are lowered during hypnosis. They therefore counsel their members to abstain from hypnosis.

Whether or not this is a legitimate position, it has certainly become a tenable view in light of the abduction/hypnotic regression stir.

Certainly we see that many forms of spiritual attack and invasion occur without the aid of hypnosis. However, the hypnotic state is induced for the specific purpose of getting through the barriers that the individual has constructed for self protection. The hypnotized person becomes "suggestible," more compliant with the will of the hypnotist. Most psychologists agree that memories can be implanted during hypnosis. This is why police must document a hypnotic retrieval session so fully if attempting to excavate a glimpse at the get-away car's license number or some other walled-in detail.

Dr. Mack, whose expertise in psychiatry included altered states of consciousness (including hypnosis), informs us that the "inducement of a nonordinary state, a modified form of hypnosis in many cases, seems to be highly effective in bringing abductees' walled off experiences into consciousness," but frankly admits that he does "not quite understand why this is so dramatically true." (*Abduction,* p. 21) He then tells us that for reasons also unknown to him, abductees seem to move more quickly into the trance state than others.

Although we lack enough evidence to make a broad indictment of hypnotic regression retrieval methods, the implications of the sessions detailed herein (which are representative of those in the literature) are strong enough to indicate that a spiritual infestation or attack occurs at the time of the hypnotic state.

Spiritual Psychology

The "Men In Black" phenomenon, referred to above, is that in which men in "official"looking apparel show up at the home or other location of a witness or abductee, demanding any and all evidence of a close encounter. They often threaten the witness if he or she ever divulges the details of the close encounter. These experiences are usually alternated with "official" looking helicopters that buzz the location of the witness. It is reported by witnesses that these MIBs are often somewhat alien looking (in the eyes), and that the helicopters have sometimes transformed themselves from disks or spheres, and back again.

Noting that the techniques of the MIBs are closely akin to spiritual brainwashing methods used by cults and Satanic groups to negatively enforce their will and control over initiates, Brad Steiger suggests that the "Brothers of the Shadow, like the MIB, are known for threatening students of the occult whenever they get too close to lifting the Veil of Isis [reaching the spiritual pinnacle]." As Madame Blavatsky says when referring to the Brothers of the Shadow, they are "the leading stars" on the great spiritual stage of "materialization."[23] The key words here are "stage" and "materialization." What Madame Blavatsky is conveying is that the occult is very "manifestation" oriented—Satan provides the curious with many signs and wonders to prove his powers. When the curious become practitioners of the occult arts, it is they who provide the uninitiated with signs and wonders. This is the great proselyting effort of Satan's kingdom. Levitating, flying, astral projection, corporeal transformation, and other manifestations of magical power are the bait. Once hooked, proselytes are controlled through psychological manipulation, such as that described by Steiger.

Negative psychological "warfare" has long been a control device of the occult, inhumane POW captors, and others with evil, dominating tendencies. As pointed out by Lewis and Shreckhise, the "aliens" appear to be playing "Good Cop/Bad Cop." Any television police drama watcher knows that this is a scenario wherein one police officer will act as the "heavy," projecting a threatening image to the targeted person. The second officer will then make an appearance, offering sympathy and hope—befriending the target—convincing the target that he is the target's friend and advocate. Manufacturing "bad" spirits or aliens drives the terrified victim into the open arms of the "good" spirits or aliens, accepting their message as beneficent, no matter how much pain is caused thereby or how at odds it may be with revealed truth. Remember that visiting spirit entities, as well as "aliens," come in the good and bad varieties by their own accounts. We know that the spirit entities are all of the same origin (Satanic), and have good reason to believe that most of the "aliens" spring from the very same source.

---

23 Brad Steiger, *Alien Meetings,* Ace Books, p. 114, quoting *New Atlantean Journal,* March 1975.

Lewis and Shreckhise further observe that the threat of superhuman retribution for pursuing the "forbidden fruit" of secret knowledge has always been too tempting for man to withstand. The threats whet the appetite, and the victim is drawn more deeply into his captor's sphere of influence.

Speaking of one woman victimized by Men In Black, Steiger notes, "At times she seemed almost to be possessed."[24] Lewis and Shreckhise speculate: "This is the purpose of MIB: to bring a person into subjection and control. By using the initial hallucinatory experience of UFO contact and subsequently the MIB threat, the contactee is subjected to a mindcontrol, brainwashing technique designed to break down his will and resistance to eventual possession."[25]

The Control System

Jacques Vallee has researched and reasoned, pondering the extraterrestrial behavior and message. He has articulated a "control" theory of his own.

> I propose there is a spiritual control system for human consciousness and that paranormal phenomena like UFOs are one of its manifestations. I cannot tell whether this control is natural and spontaneous; whether it is explainable in terms of genetics, of social psychology, or of ordinary phenomena—or if it is artificial in nature, under the power of some superhuman will. It may be entirely determined by laws that we have not yet discovered.
>
> I am led to this idea by the fact that, in every instance of UFO phenomenon I have been able to study in depth, I have found as many rational elements as absurd ones, as many that I could call friendly as I could call hostile. This is what tells me that we are working on the wrong level. And so are all of the believers, and this definitely includes the skeptics, because they believe that they can explain the facts as strongly as the most enthusiastic convert to Ms. Dixon's vision of Jupiterians![26]

Vallee explains how the system works through a deceptive reinforcement schedule:

> If the phenomenon is forcing us through a learning curve, then it has no choice but to mislead us. When Skinner designs a machine that feeds a rat only when the right

---

24 *Ibid.*, p. 117.

25 *UFO: End-Time Delusion*, pp. 137-38.

26 *Dimensions*, p. 272.

lever is depressed, this is extremely misleading for the rat. But if the rat doesn't depress the right lever, he becomes extremely hungry. Man is hungry for knowledge and power, and if there is an intelligence behind the UFOs it *must* have taken this fact into account. We also tend to forget that we have no choice either—we must eventually study UFOs, and that study, unavoidably, will in turn contribute to the reinforcement itself.[27]

Vallee suggests that it is possible to convince a major sector of our population to believe in the existence of superior extraterrestrial beings "by exposing them to a few carefully engineered scenes the details of which are adapted to the culture and symbols of a particular time and place." He then explains why he believes the UFO phenomenon/control system has been engineered.

> Could the meetings with UFO entities be designed to control our beliefs? Consider their changing character. In the United States, they appear as science fiction monsters. In South America they are sanguinary and quick to get into a fight. In France, they behave like rational, Cartesian, peace-loving tourists. The Irish Gentry, if we believe its spokesman, was an aristocratic race organized somewhat like a religious-military order. The airship pilots were strongly individualistic characters with all the features of the American farmer. . . .
>
> What is the variable being controlled in this control system? Thermostats control temperature; gyroscopes control the direction in which a rocket flies. What could a paranormal phenomenon control? *I suggest that it is human belief that is being controlled and conditioned.*[28]

We are compelled to agree with Vallee's conclusion—the UFO phenomenon is "presented" to us in a manner designed to elicit human response sufficient to manipulate human beliefs. He hesitates in his many books to forthrightly identify the source of the presentation—or at least his interpretation of it. It is significant that Dr. Vallee is a top computer scientist, specializing in the analysis and processing of scientific research data. In fact, it was while creating systems for observatories that he became interested in UFO research—as he saw UFO tracking tapes destroyed as a matter of policy. After many years of inquiry, what does Vallee conclude about the origin of the UFO phenomenon? In an interview given to the SCP Journal, Vallee revealed the following: "We believe that the thousands of cases of transformation represent one aspect of the ultimate purpose of

27 *Ibid.*, p. 275.
28 *Ibid.*, p. 276.

UFOs. They are part of a plan to deliberately move significant portions of an entire culture, or world, into acceptance or involvement in the occult, and a collective alteration in world view. This is preparatory for and necessary to the events surrounding the rise of the Antichrist."[29] It is no wonder that Vallee hesitates to share his ultimate conclusions in his books.

Does Vallee believe that UFO occupants are demonic? In explaining why the phenomenon operates as it does, he relates that some witnesses assume that they are. He says, "In fact, some witnesses have thought that they had seen demons because the creature had the unpredictability and the mischievousness associated with popular conceptions of the devil." Vallee further explains why he believes that "UFO" and "alien" encounters are so bizarre, illogical, and internally inconsistent:

> If you wanted to bypass the intelligentsia and the Church, remain undetectable to the military system, leave undisturbed the political and administrative levels of a society, and at the same time implant deep within that society far-reaching doubts concerning its basic philosophical tenets, this is exactly how you would have to act. At the same time, of course, such a process would have to provide its own explanation to make ultimate detection impossible. In other words, it would have to project an image just beyond the belief structure of the target society. It would have to disturb and reassure at the same time, exploiting both the gullibility of the zealots and the narrow-mindedness of the debunkers. This is exactly what the UFO phenomenon does.[30]

New Ideas

In describing various psychological techniques employed to open human minds to heretofore unacceptable precepts, Vallee notes further similarities between occult methods and those of UFO messengers:

> Is this confusion technique deliberately used to effect change on a major scale? Answering such questions could also help us to understand the strong resemblance that anyone who has examined the beliefs of esoteric groups could not fail to note between certain UFO encounters and the initiation rituals of secret societies. This "opening of the mind" to a new set of symbols that is reported by many witnesses is precisely what the various occult traditions also try to achieve.[31]

---

29 August 1977, p. 23.

30 *Dimensions,* p. 178.

31 *Ibid.,* p. 188.

Is this "opening of the mind" of a major portion of the population actually occurring? Yes! With the cooperation of the popular media, the "extraterrestrial savior" dream has been fostered and championed. All of our ills—pollution, energy, war, property, hunger, poverty, and social injustice—will soon be cured; just as soon as we establish permanent ties with these superior lifeforms. With the promotion of these intergalactic messiahs by the New Age movement, occult spiritualism is now centered in the cosmos—*extraterrestrials are God.*

In summing up the UFO message, Raymond Fowler concludes that the significance of their message is revolutionary. He says, "It is now time to review *what* the aliens told Betty about their *identity* and *operations.* They were brief and to the point but what they revealed will revolutionize every aspect of our lives—science, religion, philosophy, sociology—nothing will be spared." Most observers, no matter what their background or which ax they choose to grind, agree on this one point—that if nothing else, the UFO phenomenon is changing world (at least Western) beliefs in God.

Lewis and Shreckhise [Christian Fundamentalists]:

> The menace is that they lure people into an alternative philosophy and world-view that stands in direct contradiction to the Christian faith as found in the Bible.[32]

Whitley Strieber:

> The only thing now needed to make the UFO myth a new religion of remarkable scope and force is a single undeniable sighting. Such a sighting need last only a few minutes—just long enough to be thoroughly documented. It will at once invest the extraterrestrials channels, the "space brothers" believers, and the UFO cultists with the appearance of revealed truth.[33]
>
> I cannot forget my memory of the visitors' claim, "We recycle souls." It had also been said to other participants. I thought of Jo Sharp's experience, of the whole tone of what was happening to me. It was becoming clear to me that the visitors were concerned with the life of the soul as well as the body.[34]

---

32 *UFO: End-Time Delusion,* p. 20.

33 *Dimensions, Forward,* p. v-vii.

34 *Transformation,* p. 198.

Jacques Vallee:

The experience of a close encounter with a UFO is a shattering physical and mental ordeal. The trauma has effects that go far beyond what the witnesses recall consciously. New types of behavior are conditioned, and new types of beliefs are promoted. Aside from any scientific consideration, the social, political, and religious consequences of the experience are enormous if they are considered over a timespan of a generation.

Faced with the new wave of experiences of UFO contact that are described in books like *Communion* and *Intruders* and in movies like *Close Encounters of the Third Kind,* our religions seem obsolete. Our idea of the church as a social entity working within rational structures is obviously challenged by the claim of a direct communication in modern times with visible beings who seem endowed with supernatural powers.[35]

Randall Baer [former New Age Guru]:

UFO sightings and contacts have made deep inroads into the everyday fabric of much of the New Age. More than seventy-five percent of New Agers firmly believe in the existence of hosts of alien beings within and around planet Earth to help in the birthing of the New Age. It is noteworthy to point out that the strong upsurge in this trend also parallels the crystal craze and the huge increases in channeling activities in the mid-to-late 80s.[36]

Kenneth Grant [Occult Leader]:

Some believe that the UFO phenomena are part of the "miracle," a mounting mass of evidence seems to indicate that mysterious entities have been located within the Earth's ambience for countless centuries and that more and more people are being born with innate ability to see, or in some way sense their presence.

Prayer for deific intervention in ancient times has now become a *cri de coeur* to extraterrestrial or interdimensional entities, according to whether the manifestations are viewed as occurring within man's consciousness, or outside himself in apparently objective but often invisible entities. New Isis Lodge has in its possession the sigils of some of these entities. The sigils come from a *grimoire* of unknown origin which forms part of the dark quabalahs of Besqul, located by magicians in the Tunnel of Qulielfi. The *grimoire* describes Four Gates of extraterrestrial entry into, and emergence from, the known Universe.[37]

---

35 *Dimensions, Introduction,* p. xiii.

36 Randall N. Baer, *Inside the New Age Nightmare,* Huntington House, pp. 145-46.

37 Kenneth Grant, *Outside the Circles of Time,* London, Frederick Muller (1980), p. 1. Grant shares hair-raising accounts of confrontations between leaders of his group and some of these spirit beings or entities.

Brad Steiger [New Age writer]:

> Again the "angels" i.e. the space intelligences, are speaking to the prophets, the UFO contactees, in order that we might be guided through the difficult period of transition as a new world rises from the ashes of the old.[38] [Remember the Phoenix theme?]
>
> [T]hese UFO prophets have not only brought God physically to this planet, but they have created a blend of science and religion that offers a theology more applicable to modern mankind.[39]
>
> The UFO contactees may be evolving prototypes of a future evangelism. They may be heralds of a New Age religion, a blending of technology and traditional religious concepts.[40]

The Cosmic Gospel

Brad Steiger enumerates the salient points of what he terms the Outer Space Apocrypha, a distillation of the "Cosmic Gospel" being preached by contactees and channelers alike:

- Man is not alone in the solar system. He has "Space Brothers" and they have come to Earth to reach him and teach him.
- The Space Brothers have advanced information that they wish to impart to their weaker brethren. The Space Brothers want man to join an intergalactic spiritual federation.
- The Space Brothers are here to teach, to help awaken man's spirit, to help man rise to higher levels of vibration so that he may be ready to enter new dimensions.
- Man stands now in the transitional period before the dawn of a New Age. With peace, love, understanding, and brotherhood on man's part, he will see a great new era begin to dawn.
- If man should not raise his vibratory rate within a set period of time, severe Earth changes and major cataclysms will take place. Such disasters will not end the world, but shall serve as cataclysmic crucibles to burn off the dross of unreceptive humanity. Those who die in such dreadful purgings will be allowed to reincarnate on higher levels of development so that their salvation will be more readily accomplished.[41]

Steiger says that there "is an enormous amount of New Age revelatory material that has been given to the UFO contactees" and that the "Space

---

38 Brad Steiger, *The Fellowship: Spiritural Contact Between Humans and Outer Space Beings,* New York, Dolphin/Doubleday (1988), p. 194.

39 *The Fellowship,* p. 1.

40 Brad Steiger, *The Fellowship: Spiritual Contact Between Humans and Outer Space Beings,* New York, Dolphin/Doubleday (1988), p. 4.

41 *Revelation,* pp. 157-58.

beings seem very concerned with the spread of what has come to be known as New Age concepts[42]. . . ." He was told by Ox-Ho, an alien that he channeled:

> People of Earth, you are becoming fourth dimensional whether you are ready or not. Leave the old to those who cling to the old. Don't let the New Age leave you behind . . . . Earth must be cleansed. There can be no transition into a new dimension without this cleansing.
>
> The world right now is feeling the effect of the Karmic pattern of the Atlantean culture. . . [43]

Dr. Mack documents this new worldview and insight affect as well. "In November 1991, Jerry awoke, feeling a presence once again. The room was filled with an orange-red light that soon receded. The next day her mind seemed to be 'turned up full volume,' flooded with thoughts. Jerry felt as if she were filled with information of a 'universal' sort, 'soul stuff, unusual for me.' After this, as mentioned above, she wrote intensely. Her writings included a hundred poems over the next one and a half months, whereas before this 'I never wrote a poem in my life.' She found the pressure of these thoughts and writing quite overwhelming and said, 'I don't know where it's coming from'" (*Abduction,* p. 120).

"At the end of June, Dave wrote me another long letter, filled with 'strange coincidences' and new connections between the people he knows in this life and in the earlier ones. Native American spirituality, shamanism, strange powers of nature, altered realities, Chi, karate, the mastery of dreams, UFO abductions, past life experiences, and multiplicity of synchronicities are all part of a mysterious puzzle for Dave whose pieces, according to master Joe, he is learning to put together" (*Ibid.*, p. 289).

New Age ministers aboard the UFOs, and in some cases, God, dispense counsel and blessings between torture and lies. Dr. Mack records the following "blessing," performed in classic patriarchal style. "After this Joe was taken further down the hall to 'another ET' who seemed to him 'like a head honcho.' He was sitting in a chair surrounded by light which appeared to come from him. This being was taller than the others, with a face that appeared more human. 'He's putting his hands on my head. It's like he's baptizing me. He likes me—he's energizing me. He's blessing me. He's

42 *The Fellowship,* pp. 39-40.

43 *Revelation,* p. 204.

giving me something to help me hang in there . . . He's giving me strength and knowing, just knowing that I'm not alone. I'm loved, and I'm connectable' with both them and on Earth." (*Ibid.*, p. 182)

Vallee discusses the "new" religious and other belief systems being adopted (and promoted in some cases) by contactees and many who read about their encounters, and offers his analysis of the results of the "control system" being imposed on us. He observes that in the same manner that science has gradually undermined faith in religion, now, faith in science is deteriorating with the advent of the New Age "science-spiritualism" being heralded by the UFO occupants.[44] This new belief system disregards the revealed truth of religion and the discovered truth of scientific experimentation and observation, leaving believers with the mysticism of the East and the dogma of the Dark Ages—centered in telepathic alien humanoids who claim to be the Father of Christ.

Are people with strong Christian backgrounds easily duped into accepting this new religion? Many find that the UFO message is "compatible" with God's Plan. They believe that the extraterrestrials are part of His plan to usher in the Millennium. The UFOs are the signs and wonders in the heavens foretold by the prophets they say. The extraterrestrials will direct the last-day events from their spacecraft. They might even be the angels who will sound the trumps, according to some.

Whitley Strieber, who like many others has been plagued by the visitors or entities throughout his life, observes: "Never, in those bleak April days, could I have imagined the subtlety of the plan that they were carrying out. Nor could I have seen the magnificent brilliance of the mind behind it."[45] Yes, Strieber, who considers himself to be of high intellectual stock, is overwhelmed by the "power and intelligence" of the visitors—he too believes it to be "more than human." He perceives that the beings have a plan of their own, and that the "One" behind it is brilliant in his execution of The Plan. Strieber worries:

> If I had been having these encounters throughout my life, then what had I become? Why were my visitors so secretive, hiding themselves behind my consciousness? I could only conclude that they were using me and did not want me to know why.
>
> Frankly, I found this idea deeply disturbing. What were the visitors' motives? *Communion* had become a number-one bestseller. What if they were dangerous? Then

---

44 Jacques Vallee, *Messengers of Deception,* pp. 221-23.

45 *Transformation,* p. 55.

> I was terribly dangerous because I was playing a role in acclimatizing people to them. And if they were benevolent? Then the agonizingly difficult task of bearing witness to their reality would turn out to be worthwhile.
>
> My desperation increased as I searched across the years of my past, seeking answers.[46]

This is the sense that many abductees have of the abduction phenomenon. Dr. Mack reports that Scott, too, felt that this was the blueprint. "At our support group meeting on February 24, Scott, just back from Los Angeles, spoke of the feeling that we are being 'prepared for' something, that there was perhaps a 'plan' of some sort, that we are not in control and 'somebody else' is 'running the show . . . Getting through the trauma part,' he said, 'has opened up the real stuff, the spiritual behind it,' and he spoke further of a 'bigger power' at work in his experiences" (*Abduction,* p. 97).

---

46 *Transformation,* p. 96.

Conclusion

The encounter phenomenon appears to be part of a great plan—The Plan—and it seems to be juxtaposed to God's Plan. The details of the abductions and the elements of the alien message point to Satanic efforts as the probable source of the alien abduction/message phenomenon. Is this possible? We must ask ourselves this because, as Whitley Strieber points out, we are modern people, well-educated and too sophisticated to set fires to drive away the evil spirits that inhabited the rural countrysides of millennia past. Yet, the evidence tells us that the current phenomenon mimics the historical phenomenon in every way.

Many investigators and writers are philosophizing about the origin and nature of UFOs and their occupants. More and more we read that there is something "evil" about them—something "inhuman." This is the apparent, even obvious, conclusion being reached by researchers. But many abductees, of all people, are telling us that even though they were brutalized and violated by the entities, they feel that no malice is intended, that the beings are benevolent. Many tell us that the UFO occupants are just so advanced that they have difficulty communicating effectively with us at our level—much as a zoologist inadvertently offends a chimpanzee when attempting to communicate or teach a lesson. Extraterrestrial proponents would have us believe that the entities "act out" their message and fill our minds with mythical symbolism in an effort to convey their higher message in a universal language. Is this really what the "aliens" are up to? The author does not think so.

Our first consideration is the multiplicity of alien types. We have presented to us a scenario in which there appear to be good aliens and bad aliens. Even the aliens make this assertion. First, the gnomes, or small greys. They profess love at times, but demonstrate unpredictability and rage at others. They lie, kidnap, blaspheme, mislead, inflict pain and suffering, exploit, torment, invade, steal, control, mutilate, impregnate, take babies, frighten children, teach reincarnation, and kill. However, they do now and again warn us about the dangers of nuclear energy. Reflect now on the angelic human aliens. They seem benevolent, caring, and superior in every way. Yet, they appear to be associated with the gnomes and indulge in, and even direct, the activities of the small greys just enumerated above.

In Western philosophy, Jesus taught that we would "know them by their fruits" (Matthew 7:16). He explained, "Even so every good tree bringeth forth good fruit; but a corrupt tree bringeth forth evil fruit" (Matthew 7:17). By analogy, the fruit of these "aliens" is corrupt and evil. The cries of those who say that we need to give the misunderstood aliens a chance are the dysfunctional chants of enabling victims. Those who claim that the "aliens" are establishing a symbiotic relationship with our race ignore the unmistakable indicators that they are, in fact, imposing a parasitic enslavement on us. This is the appearance of their design, anyway. If we were to put our very best brains to the task of procuring the voluntary enslavement of the world's population, the elements present in the UFO phenomenon would be essential. Superior visitors from afar, bearing higher technology has always been a workable rouse—Earth's history is replete with examples. The only thing these brains would lack to fulfill their design is the technology. Whoever the "aliens" are, they have the technology—real or perceived. Either way, it is sufficient to entrap a world hungry for an easier, *Star Trek* lifestyle.

If the gnomes and angelic humans are evil, then what about the huans—those who act like courteous, noninterfering explorers? These are the type encountered by Udo Wartena. These profess to be extraterrestrial, yet act very much like well-mannered, educated humans. They explain their purposes and their methods. They obey law and ethics. They speak with their mouths and shake with their hands. They have no message, but would happily speak of philosophical concepts acceptable to Western culture, including religion, if they were permitted. Are these any different, any more feasible or trustworthy than the superior humans that accompany the gnomes? We cannot know, but their fruits appear good—so far, at least.

Whatever its source, the "alien" presence *will* continue to escalate. Whether the aliens are demonic or not, demonic hordes *will* utilize the concept, in part at least, to help establish Satan's premillennial kingdom. This we know from the many "alien" contacts already being made through New Age channelers.

Will there be a large scale invasion by bad aliens, or a war between good and bad aliens locked in battle over our planet? Will there be emissaries from afar, or supermen from our ranks, enlightened by the intergalactic mind? What drama awaits us in the final scene? However the drama unfolds, it is only that—theater.

It matters little *how* Satan deceives the nations—prophetically, we know he will. Beware of the Prince of the power of the air and of false messiahs and false christs, coming in power and glory and invoking the holy names of God. Beware of false prophets and superhumans who work miracles in our sight and console us with dispensations of forgiveness and extensions of time. These will come, speaking in dignified tones and reasoning with superior intellects. They will offer hope and answers to a world poised on the precipice of certain destruction. They, and they alone will present workable solutions to seemingly insurmountable problems. Their price? We must merely "surrender the illusion of control." That is all they require.

Whether as aliens, superhumans, or ascended spirits, Satan's servants will come working miracles and ensnaring the world. Viewing these times remotely, the Apostle Paul gave us comfort, calling us the children of light, and reassured us that we need not be deceived if we will exercise our God-given gifts and reject the great lie.

> But of the times and the seasons, brethren, ye have no need that I write unto you. For yourselves know perfectly that the day of the Lord so cometh as a thief in the night. For when they shall say, Peace and safety; then sudden destruction cometh upon them, as travail upon a woman with child; and they shall not escape. But ye, brethren, are not in darkness, that that day should overtake you as a thief. Ye are the children of light, and the children of the day: we are not of the night, nor of darkness. Therefore let us not sleep, as do others; but let us watch and be sober [1 Thessalonians 5:1-6].

The Emperor has no clothes.

Index

A

Adamski, George, 124.
Airship, 37, 56-57.
Angel(ic), 22, 23, 29, 31-34, 36, 114, 133, 134, 135, 150, 226, 253.
Angelucci, Orfeo, 125.
Animals, 12, 26, 28, 65, 58, 71, 75-76, 80, 85, 98, 100, 105, 107, 112, 123, 124, 125, 129, 130, 131, 133, 134, 138, 148, 157-158, 176, 182, 184, 194, 204, 205, 209, 228, 231, 239, 244, 246, 250, 253, 263, 266.
Antichrist, 152, 156, 165, 225, 248.
Arnold, Kenneth, 40.

B

Bedroom visitation, 98, 139, 175, 190.
Bible, 25-31, 139, 159-67, 214.
Bigelow Holding Company, 13.
Blumrich, Josef F., 30.

C

Carter, Jimmy, President, 72-73.
Channeler, medium, 76, 127, 152, 162-65, 233, 252.
Child(ren), baby(ies), 17, 31-33, 50, 59-60, 80, 89, 92-94, 96, 109, 122, 123, 129, 131-132, 139-144, 149, 150, 157, 158, 160, 167, 170, 184-87, 189, 191, 193, 194, 196, 197, 201, 202-204, 217, 218, 224, 226, 229, 233-35, 240, 241, 265, 250, 262, 263, 267.

D

Davis, Kathy (Debra Tomey), 55, 63, 80, 91, 98, 221.
Davis, Laura, 94, 221.
Devil, Satan, 14, 55, 152, 157-60, 175, 207-54.
Disasters, 18, 21, 22, 125, 138-39, 150, 153, 222, 226, 251-52.
Dream, 14, 97, 173, 175, 234.

E

Ezekiel, 28-31.

F

Fairy, 175-83, 230-31.
Fatima, Portugal, 184-85.
Foofighters, 39.
Fowler, Raymond, 53, 69, 128, 142, 172, 182, 242, 249.

G

Gallup, poll, 151.
Ghost, 14, 16, 115, 174, 175, 187-88, 214-216, 220.
Giants, sons of heaven, watchers, 31, 33.
God, 27, 34, 36, 112, 115, 133-136, 148, 207, 214, 219, 226, 228, 231, 234.
Gun, 78, 79, 80-81, 106, 111-113, 202.

H

Hill, Barney and Betty, 87-89, 94.
Hoax, 19, 37, 41, 42, 62-63, 70-72, 165, 221, 233.
Hopkins, Budd, 13, 55, 63, 89, 91, 95, 96, 242.
Hubbard, Barbara Marx, 150.
Hybrid, 140-43, 221.
Hypnosis, 94, 133, 242-44.

I

Implant, 96, 134.

K

Kelley, Kentucky, 80.

L

Landing site, 62-65, 199.
Lazar, Robert, 66-69.
Love, 82, 93, 109, 126, 136, 142, 203, 236.
Luca, Betty (Andeasson), 56, 76, 88, 128, 145, 217, 242.

M

Mack, John E., M.D., 13, 21, 137, 161, 173, 191-96, 236, 241.
Madonna (Virgin Mary), 55, 184-86, 208.
Master, the One, 115, 124, 129-136, 149, 152, 159, 253.
Men in Black (MIB), 52-53, 244-45.
Monster, 14.
Montgomery, Ruth, 150, 190, 232.
Moon, 124.

N

New Age, 125, 132, 146-156, 236, 249-250, 253.
Newhouse, Delbert, 43, 71.

O

Ouija board, 186.
Old age, longevity, 76, 82, 203.
Out-of-body experience (OBE), 14, 131, 172-74, 195.

P

Photograph, 42-49, 65, 79, 189, 199-201.
Plan, The, 150, 255.
Polygraph, 44-45, 103.

R

Reincarnation, 161, 191-95.
Religion, 21, 38, 115, 128, 132, 136, 137, 146-167, 175, 184-86, 218, 249, 250-51.

Roper Organization, 12-17, 20, 84, 98, 215.
Roswell, New Mexico, 65.

S

Salisbury, Frank B. Ph.D. 169, 186, 209.
Scar, 14, 18.
SETI, 72.
Sex, 31, 32, 89-90, 111, 116, 193, 216, 224, 239.
Smith, Wilbert b., 70.
Steiger, Brad, 167, 211, 233, 245, 251.
Strieber, Whitley, 22, 70-71, 174-76, 179, 187, 218, 229, 236-40, 249, 253.
Subterranean, 75-76, 130.
Superhuman, 224-26.
Survey, 12-17, 20, 84.

T

Tunguska, Russian, 73-74.

V

Vallee, Jacques, 35, 167, 176, 181-82, 184, 230, 246-49, 250.
Villas-Boas, Antonio, 90.
Voice, 47-52, 98, 106, 114, 126, 129, 135, 136, 183, 218.

W

Walters, Ed, 44-51, 76, 78, 95, 143, 199, 216.
Walton, Travis, 100-03.
Wand, hand-held device, 31, 78, 79, 179, 236.
Wartena, Udo, 81-83, 143, 202.
Water, 82, 97, 99.

Y

Year 2,000, 137, 149.

Z

Zeta Reticuli, 68.